INDUSTRIAL ROBOTS PROGRAMMING:
BUILDING APPLICATIONS FOR THE FACTORIES OF THE FUTURE

INDUSTRIAL ROBOTS PROGRAMMING:
BUILDING APPLICATIONS FOR THE FACTORIES
OF THE FUTURE

J. Norberto Pires
Mechanical Engineering Department
University of Coimbra, Portugal

 Springer

J. Norberto Pires
Mechanical Engineering Department
University of Coimbra
Portugal

Industrial Robots Programming: Building Applications for the Factories of the Future

Library of Congress Control Number: 2006932582

ISBN 0-387-23325-3   e-ISBN 0-387-23326-1
ISBN 9780387233253

Dedicated to the memory of my father Joaquim
and to Dina, Rita, Beatriz and Olímpia.

Foreword

Robots have traditionally been used to work in industrial environments, as they constitute the most flexible existing automation technology. In the recent years, manufacturing systems are becoming more autonomous requiring less operator intervention and a higher degree of customization and reconfigurability for disparate applications. In this scenario, robot programming is a key factor toward building the applications for the factories of the future.

This book by J. Norberto Pires constitutes a unique and authoritative reference in our professional field, as one of the very few books written by an academic with a strong industrial cut. The focus is on the software interfaces enabling humans and machines to effectively cooperate on the shopfloor. Several sensors and controllers are analyzed in detail, leading to the realization of interface devices using e.g. speech recognition and CAD models, and their versatility for a number of industrial manufacturing systems is enlightened.

Easy to read, rich in worked out examples and case studies, the book is complemented with additional experimental material available on a web site, including code and multimedia files, which the author promises to update regularly.

It is my conviction the book will be appreciated by a wide readership, ranging from technical engineers wishing to learn the foundations of industrial robotics to scholars and researchers wishing to understand the needs and the potential of a new generation of advanced industrial robots to be developed in the next decade.

Bruno Siciliano
Professor of Control and Robotics at the University of Naples
President-Elect of the IEEE Robotics and Automation Society

Preface

A scientific and technical book is a starting point. A source of information for people browsing for details, a guide for others trying to build similar or related solutions, or a source of inspiration for yet others wondering about how things work.

This book was written by an engineer and university professor which has been active in the field of industrial robotics since 1994. It was planned, designed and built to serve engineers looking for better and more efficient systems, but also to serve academic readers interested in the robotics area. The book focus mainly on industrial robot programming in the beginning of the twentieth first century, namely on the important issues related with designing, building and operating flexible and agile robotic systems. It explores in detail the issue of software interfaces, but also input/output devices and several industrial and laboratory examples. In fact, the book uses several types of fully worked out examples to illustrate and clarify concepts and ideas, enabling the reader to see them working and even to test some of them. Most of the experimental material used in this book can be obtained from:

http://robotics.dem.uc.pt/indrobprog

This site will be updated regularly by the author constituting a source of information, code and multimedia files which complement the contents of the book.

Finally, the author wants to thank deeply to all the persons that contributed to this book, namely all his undergraduate and graduate students, specially his Ph.D. students Tiago Godinho and Germano Veiga, and his M.Sc. student Ricardo Araújo for their help and support in building and testing some of the solutions presented in the book.

J. Norberto Pires, Coimbra, Portugal, 2006

Contents

1

Introduction to the Industrial Robotics World

1.1 Introduction

Robotics is a subject that leaves nobody indifferent. No matter if they are used to work in industry or at our homes, mimic some of the human capabilities, or used to access dangerous environments, launched to space, or simply used to play with, robots are always a source of interest and admiration. Here the focus is in robots used to work on industrial environments [1], i.e., robots built to substitute man on certain industrial manufacturing tasks being a mechatronic coworker for humans.

In fact, actual manufacturing setups rely increasingly on technology. It is common to have all sources of equipment on the shop floor commanded by industrial computers or PLCs connected by an industrial network to other factory resources. Also, manufacturing systems are becoming more autonomous, requiring less operator intervention in daily operations. This is a consequence of today's market conditions, characterized by global competition, a strong pressure for better quality at lower prices, and products defined in part by the end-user. This means producing in small batches, never risking long stocks, and working to satisfy existing customer orders. Consequently, concepts like flexibility and agility are fundamental in actual manufacturing plants, requiring much more from the systems used on the shop floor. Flexible manufacturing systems take advantage of being composed by programmable equipment to implement most of its characteristics, which are supported by reconfigurable mechanical parts.

Industrial robots are good examples of flexible manufacturing systems. Using robots in actual manufacturing platforms is, therefore, a decision to improve flexibility and to increase the agility of the manufacturing process. If the manufacturing processes are complex, with a low cycle time, and have a lot of parameterization due to the diversity of products, then using robots is the correct decision, although it isn't enough for a complete solution. In fact, engineers need to integrate other technologies with the objective of extracting from robots the flexibility they can offer. That means using computers for controlling and supervising manufacturing systems, industrial networks, and distributed software architectures [2,3]. It also means designing application software that is really distributed on the shop floor, taking advantage of the flexibility installed by using programmable equipment. Finally, it means taking special care of the human-machine interfaces (HMI), i.e., the devices, interfaces, and systems that enable humans and machines to cooperate on the shop floor as coworkers, taking advantage of each other's capabilities.

1.2 A Brief History of the Industrial Robot

The word "*robot*" comes from the Czech "*robota*" which means tireless work It was first used in 1921 by the novelist *Karel Capek* in his novel "*Rossum's Universal Robots*". Capek's robots (Figure 1.1) are tireless working machines that looked like humans ,and had advanced capabilities even when compared with actual robots. The fantasy associated with robotics offered by science fiction movies, and printed and animated cartoons is so far from reality that actual industrial robots seem primitive compared with the likes of *C3PO* and *R2-D2* (from the movie *Star Wars*), *Cyberdyne T1000* (from the movie *Terminator II*) *Bishop* (from the movie *Alien II*) and *Sonny* (from the movie *I Robot*), for example.

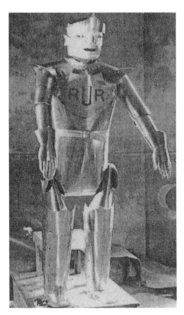

Figure 1.1 A robot from *Karel Capek's* novel "*Rossum's Universal Robots*"

But robotics was a special concern of the most brilliant minds of our common history, since many of them took time to imagine, design, and build machines that could mimic some human capabilities. It is one of the biggest dreams of man, to build obedient and tireless machines, capable of doing man's boring and repetitive work; an idea very well explained by *Nicola Tesla* in his diary [4]:

> *"... I conceived the idea of constructing an automaton which would mechanically represent me, and which would respond, as I do myself, but, of course, in a much more primitive manner, to external influences. Such an automaton evidently had to have motive power, organs for locomotion, directive organs, and one or more sensitive organs so adapted as to be excited by external stimuli ...".*

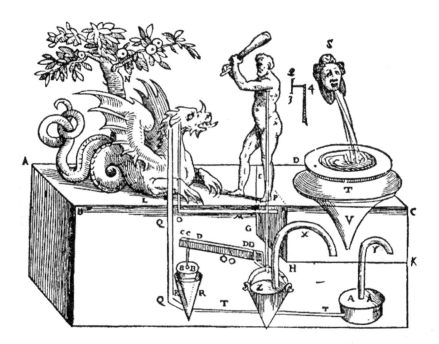

Figure 1.2 Water clocks designed by *Ctecibius* (270 B.C.)

Today's challenge is to consider robots as human coworkers and companions, extending human capabilities to achieve more efficient manufacturing and to increase the quality of our lives. This book focuses on industrial robotic coworkers. The fields of robotics that consider the companion aspects, namely service robotics and humanoid robotics, are not covered in this book. Nevertheless, the social perspective of using robots not only as coworkers, but also as personal assistants, is very promising. In fact, due to several social and economical factors, we are required to work until very late in life: It is common in Europe to only allow retirement when a person is near seventy years old. Since our physical and mental capabilities decrease with time, the possibility of having mechanical assistants that could help us in our normal routine has some valuable interest.

Robotics can be traced back to 350 B.C., in the ancient *Greece*, to the fabulous philosopher and mathematician *Archytas of Tarentum* (428-347 B.C.) and a demonstration he made in front of the *metropolis* senators. A strange machine that he called "*the pigeon*" was capable of flying more the 200m, using some type of jet propulsion based on steam and compressed air: a great achievement for the time (the invention of the screw and also the pulley are attributed to *Archytas*).

Figure 1.3 A Greek design adapted by *al-Jazari* for a garden hand-washer

In 270 B.C., also in ancient *Greece*, the civil engineer *Ctecibius* was capable of building water clocks with moving parts (Figure 1.2). His work had followers like *Phylo of Byzantium* author of the book "*Mechanical Collection*" (200 B.C.), and

Hero of Alexandria (85 B.C.), *and Marcus Vitruvius* (25 B.C.). In the twelfth century, the Arabian *Badias-zaman al-Jazari* (1150-1220) recollected some of the Greek developments in the book *"The Science of the Ingenious Devices"* [5] (Figure 1.3), and that is how they reached our time. In those early times the problem was about mechanics, about how to generate and transmit motion. So it was mainly about mechanisms, ingenious mechanical devices [5,6].

Then in the fifteenth century, *Leonardo da Vinci* showed indirectly that the problems were the lack of precision and the lack of a permanent power source. He designed mechanisms to generate and transmit motion, and even some ways to store small amounts of mechanical energy [7]. But he didn't have the means to build those mechanisms with enough precision and there was no permanent power source available (pneumatic, hydraulic, or electric). Maybe that was why he didn't finish his robot project [5,6], a fifteenth century knight robot (Figure 1.4) intended to be placed in the *"Salle delle Asse"* of the *Sforza* family castle in Milan, Italy. It wasn't good enough. Or it was so revolutionary an idea for the time that he thought that maybe it was better to make it disappear [5,6].

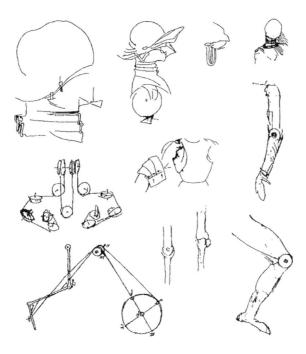

Figure 1.4 Leonardo's studies for a humanoid robot

And then there was the contribution of *Nicola Tesla* at the turn of the nineteenth century. He thought of using *Henrich Hertz's* discovery of radio waves (following the work of *James Clerk Maxwell* about electromagnetic phenomena) to command an automata. He built one (Figure 1.5) to demonstrate his ideas and presented it in New York's *Madison Square Garden* in 1898 [4,6]. The problem then was that machine intelligence was missing. Robots should be able to do pre-programmed operations, and show some degree of autonomy in order to perform the desired tasks. When that became available, robots developed rapidly, and the first industrial one appeared in the early 1970s and spawned a multi-million dollar business.

After that, robotic evolution was not as fantastic as it could have been, since there was a lot to do and the available machines were sufficiently powerful to handle the requested jobs. Manufacturers were more or less happy with their robots, and consequently industrial robots remained position-controlled, somehow difficult to program by regular operators, and really not especially exciting machines. Features currently common in research laboratories hadn't reached industry yet because of a lack of interest from robot manufacturers. Nevertheless, there was a considerable evolution that can be summarized as follows.

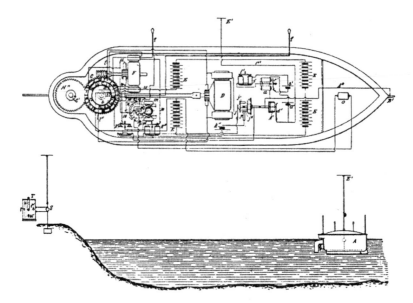

Figure 1.5 Nicola Tesla's remote-controlled miniature submarine

In 1974, the first electrical drive trains were available to use as actuators for robot joints. In the same year, the first microprocessor-controlled robots were also available commercially.

Around 1982, things like Cartesian interpolation for path planning were available in robot controllers, and many of them were also capable of communicating with other computer systems using serial and parallel interfaces. In the same year, some manufacturers introduced joystick control for easier programming, and the *teach pendant* menu interface.

In 1984, vision guidance was introduced as a general feature for tracking, parts identification, and so on.

In 1986, the first digital control loops were implemented enabling better actuator control and enabling the use of AC drives.

Networking is a feature of the 1990s, with several manufacturers implementing networking capabilities and protocols.

In 1991, there was the implementation of digital torque control loops, which enabled, for example, the utilization of full dynamical models; a feature only available in the first robots around 1994.

During the period 1992-1994 several manufacturers introduced features like Windows-based graphical interfaces, virtual robot environments for off-line programming, and *fieldbuses*.

Robot cooperation is a feature introduced from 1995 to 1996.

Figure 1.6 Actual robot manipulators

Around 1998, robot manufacturers started introducing collision detection to avoid damaging robots, and load identification to optimize robot performance. Since then other features include fast pick and place, weight reduction, optimized programming languages, object-oriented programming, remote interfaces using RPC sockets and TCP/IP sockets, *etc.*. Figure 1.6 shows some of the robot manipulators available currently on the market.

So how do we define robotics then? Is it a science? Is it a technique or collection of techniques? If the reader opens a robotics book something like this appears:

> *"A robot is a re-programmable multi-functional manipulator designed to move materials, parts, tools, or specialized devices, through variable programmed motions for the performance of a variety of tasks", from the book Robotics – Control, Sensing, Vision and Intelligence, Fu, Gonzalez, Lee, MacGraw Hill, 1987.*

Although correct, despite being restricted to robot manipulators, this definition doesn't give the correct idea. The common sense image of a robot is usually associated with strong and superb machines, tireless (like *Karel Capek's* machines), obedient (*"yes, noberto san ..."*), but nevertheless, fascinating machines that make us dream. And that fascination is not in that definition.

As with everything, we should look to the past and pick what was fundamental for the history of robotics in terms of ideas and dreams. From the *Greeks* and *Arabs* we should pick their idea of *"ingenious devices"*. In fact, robotics is very much about mechanics, motion, mechanisms to transmit motion, and having the art and the skill to design and build those mechanisms. Yes, "ingenious devices" is really a good start.

Then we should turn to *Leonardo* (sixteenth century) and look to his quest for *"... precision ..."* and *"...permanent power source ..."*. He understood that robots need parts built with very high precision and a permanent power source. That was not available at his time, *i.e.*, machine tools and a permanent power source (electric, hydraulic, or pneumatic).

Finally, we should read *Nicola Tesla* and observe his outstanding and visionary work. He understood that robots are a consequence of dreams and neat ideas. Robots need to be controlled and programmed, distinguish situations, *etc.*, have ways of *"understanding"*, and that means using computers, electronics, software, and sensors, in a way to enable machines to be programmed and to sense their environment. Those are the elements that enable us scientists, engineers, and robot users to try different things and new ideas, being a source of fascination. In his own words [4]:

> *"... But this element I could easily embody in it by conveying to it my own intelligence, my own understanding. So this invention was evolved, and so a new art came into existence, for which the name "teleautomatics" has been suggested, which means the art of controlling movements and operations of distant automatons.*

Therefore, we can define robotics as a science of generic, ingenious, precise, mechatronic devices, powered by a permanent power source; a science that is open to new ideas and that stimulates the imagination. A stimulus so strong that it attracted many of the best minds of our common history, *i.e.*, authors of the work that constitutes the legacy that we humans leave for the future.

1.3 Using Robotics to Learn

Putting robots in space, and in other planets, is a very exciting field of modern robotics. This and other fantastic achievements justify the enormous interest about robots and robotic applications. Only a few engineering fields are as multidisciplinary as robotics, i.e., areas that require knowledge of as many different scientific and technical disciplines. Robotics integrates an extensive knowledge of physics, mechanics, electronics, computer science, data communications, and many other scientific and technical areas important for the design, construction, and operation of machines that execute human-like functions.

Figure 1.7 Robot MER-A (*Spirit*) sent to Mars in June of 2003 (from NASA) [8]

In this section a small mobile robot, named *Nicola*, is presented. The robot is constructed, using commonly available industrial equipment, to be commanded from a typical personal computer running standard operating systems and software development tools. The final goal is to demonstrate what is involved in the construction of this type of robot, showing that it is possible to play with science and technology and in the process learn and spend a fantastic time. The robot *Nicola* will be presented step-by-step with enough detail for showing what is involved.

NASA initiated in June 2003 a new mission to further explore Mars, the great red planet of our solar system [8]. The allure of Mars is based on its proximity to Earth, but also on the assumption that it was once like Earth, with water available on the surface and life, before changing gradually to a hot and dusty planet. In this mission, NASA used again semi-autonomous mobile robots to explore the planet. These Mars exploration rovers (MER – Figure 1.7), named *Spirit* and *Opportunity*, are capable of navigating the surface of the planet, analyzing rocks and land, and sending back pictures, videos, and results from experiments carried out on the planet's surface. The spaceship that carried *Spirit* was launched on June 10, 2003, and arrived on Mars on January 4, 2004. In turn, the spaceship that carried *Opportunity* left on July 7, 2003, and arrived on Mars on January 25, 2004.

The utilization of these robots was also a dream of the great Croatian inventor *Nicola Tesla* (1845-1943), a man that gave a pioneering and visionary contribution for the evolution of robotics. He worked with the legendary *Thomas Edison* and was a tireless, dedicated, and bright inventor. *Tesla* was the archetype of the inventor: solitary, absent minded, abstracted of the normal things of life, with an exclusive dedication to his work and visionary. At the end of the nineteenth century he dreamt (doesn't everything begins like this?!) of *automatons* capable of performing tasks only possible to intelligent living creatures. For that, the *automaton* needed an element equivalent to the human brain. Since that seemed complicated, he thought about using his own brain for commanding the automaton [4].

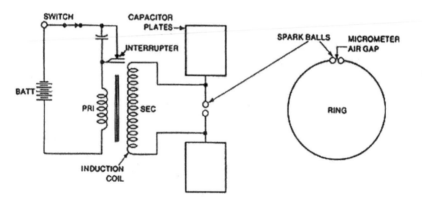

Figure 1.8 Heinrich Hertz's first transmitter, 1886 schematic

That capacity of commanding distant automatons was achieved using *Henrich Hertz* waves (published in 1887 in a treatise named "*Optice Elettrica*"). *Tesla* had access to *Hertz's* publications and saw in his radio transmitters and receivers (Figure 1.8) a way to implement his own ideas. To demonstrate the principle, *Tesla* built a model of a submarine (Figure 1.5) controlled remotely using coded hertz impulses (controlled by radio, therefore). He could command the boat to turn to the right or to the left, submerge and emerge, *etc*. Despite the enormous interest of the new invention, which he demonstrated in the *Madison Square Garden* of *New York* City (1898), before an overwhelmed audience, he failed to obtain support to continue his efforts on the subject.

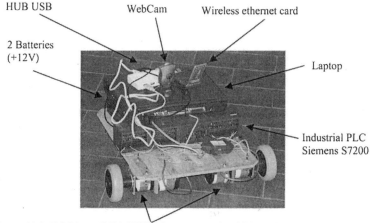

HUB USB   WebCam   Wireless ethernet card

2 Batteries
(+12V)

Laptop

Industrial PLC
Siemens S7200

Power Unit: DC Motor (24V, 50W), planetary gearhead 25:1, velocity control circuitry

a)

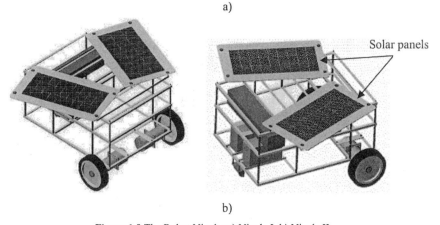

Solar panels

b)

Figure 1.9 The Robot Nicola: a) Nicola I; b) Nicola II

But it was a fabulous advancement for the time (nineteenth century). How would it be building a system with those characteristics today? Using common industrial equipment, wireless communications, actual operating systems, and well known programming tools?

That is the goal of our robot *Nicola*, *i.e.*, to show that *Tesla's* dream is still actual, and that despite the sophistication of those robotic space explorers (Figure 1.8), the technology involved and the concepts are simple, accessible, and fun to learn how it all basically works.

1.3.1 Constitution of the Robot Nicola

The robot *Nicola* is very simple. Basically it is a three-wheel robot with two power wheels in front and a free wheel in the back (Figure 1.9). The wheels used are of the same type that can be found in office chairs and other office equipment. Each of the two power wheels are equipped with a power unit composed of:

1. One 24 V DC motor (máx. power 50 W, máx. velocity 3650 rpm, máx. torque 0.17 Nm), model MDLC-58 from *Maclennan Ltd.* [9]
2. One 25:1 gear unit, model IP57-M2 from *Maclennan Ltd.* [9]

The selected DC motor is equipped with a velocity control loop (Figure 1.10), which makes it very simple to linearly control velocity just by feeding the unit with a 0-5 V analog signal. The control circuit is a very simple electronic circuit composed of a velocity control loop and a power amplifier. The velocity control loop makes the motor velocity proportional to the commanding analog signal (0-5 V in magnitude), and the rotating velocity is defined by a digital input (0 – positive direction, 1 - negative direction).

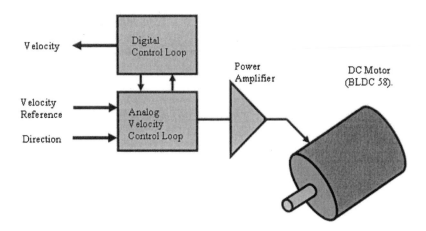

Figure 1.10 Diagram of the velocity control circuitry [9]

Using this power unit, attached to each wheel, there is no need for a mechanical steering mechanism since the electric differential effect can be used to steer the robot, i.e., varying the speed of each independently wheel it is possible to turn to the right and to the left with high-precision and several curvature radius. For example, if the speed of the left wheel (v_l) is equal to the speed of the right wheel (v_r), the robot moves forward in a straight line ($v_l = v_r > 0$). If we change the sense of rotation of the wheels ($v_l = v_r < 0$), the robot moves backwards also in a straight line. Making $v_l > v_r$, the robot turns to the right, and with $v_l < v_r$ it turns to the left. Adjusting the value of v_l and v_r several curvature radius may be obtained. Finally, making $v_l = -v_r$ the robot turns about itself.

Furthermore, with the objective of using industrial components, the robot uses a medium class PLC (*Programmable Logic Controller*) to interface with sensors and actuators. The selected PLC is a *Siemens S7-200* (DC model with the 215 CPU), equipped with a 12-bit resolution analog module (module EM235, with three inputs and one output) [10].

To command the robot, a laptop is installed on the robot, connected to the PLC using a serial link (RS-232C channel). The software running on the laptop was built to work as a *TCP/IP socket server*, enabling commands by any authorized remote client. The operating system running on the PC is the *Microsoft Windows XP*, which makes it easy to add advanced services, attach devices (like network devices, Webcams, etc.), and explore them from available software developing tools (*Visual Basic, C++, C#*, etc.).

1.3.2 Nicola Software

The software designed to control and operate Nicola is divided into three levels, identified with the programmable hardware components that constitute the robot:

1. The PLC that implements the low-level interface with sensors and actuators
2. The on-board PC used to manage the robot and interface with remote users
3. The commanding PC, i.e., the computer used to command the robot and monitor its operation

In the following sections the software will be presented in detail. The interested reader can download the source code from [11].

1.3.2.1 PLC Software

The mission of the PLC is to interface with analog and digital sensors that could be used with the robot, and to control the two DC motors that move the robot and any other actuator that could be added to the system. Consequently, a PLC is a good choice since this is basically what is required from them in industry, i.e., to work as local and low-level interfaces with sensors and actuators implementing sequential commanding routines. In addition, PLCs are very easy to use and to program, which also justifies the solution. The only difficulty with the PLC is the need to have it working as a server, executing the commands sent by the on-board PC that manages the robot (Figure 1.11). This means that the PLC should implement the services required to operate the robot, namely:

1. The possibility to change any analog or digital output
2. The possibility to access any analog or digital input
3. The possibility to command macros, or batches of functions

4. The possibility to receive events with detailed information about the status of the robot.

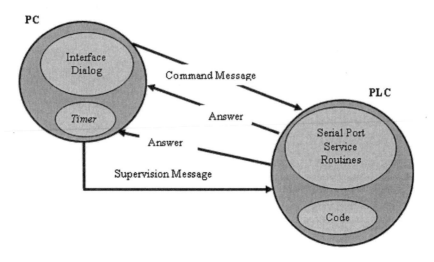

Figure 1.11 Messages between the on-board PC and the PLC

Table 1.1 List of PLC command codes

Command	Parameter 1	Parameter 2	Description
159	120 + output	Valor	Changes the specified analog output.
160	120 + input	-	Reads the actual value of the specified analog input.
200	120 + output 130 + output		Activates the specified digital output of the first output block (120) or of the second output block (130).
201	120 + output 130 + output		Deactivates the specified digital output of the first output block (120) or of the second output block (130).
253	-	-	Supervision message.

This idea is very simple and not different from what is done in more advanced machines, like the industrial robots. From the remote client, properly authorized, the user accesses memory zones to change some predefined variables (bytes, word or double-word variables). If the value of those variables is used in the programmed instructions, it is possible to execute only the intended sequences just by comprehensively changing the values of those variables. The PLC answers to remote commands sent with a pre-defined format and with a maximum length of

100 bytes. The first byte of the commanding message specifies the command, and the following bytes are parameters (see Table 1.1).

The synchronous answer of any command is a copy of the message received, which enables the commanding PC to check if the command was well received using for example an ACK-NACK (acknowledge – not acknowledge) protocol. Besides that, there is a special command (code = 253) used for monitoring the PLC state. When the PLC receives this command it should answer by sending the state of all its IO inputs and outputs. This message should be sent frequently to track the robot state. In the robot Nicola this message is associated to a 500 ms timer, which means that the robot state is updated at a frequency of 2 Hz.

Any asynchronous answer contains the execution results of one command. For easy identification from the serial port interrupt routine, the first byte of the answer identifies the code of the executed command. The user commands should be associated with user actions like pressing software buttons or keyboard buttons, etc. When the PLC receives a command, it transfers the received data into a pre-defined memory zone starting with register VB90. Consequently, if the command contains n bytes, with $n <= 100$, the following happens:

>Byte **VB90** – contains the number of byte received
>Byte **VB91** – contains the character (code) that identifies the command
>Byte **VB92** – contains parameter 1
>...
>Byte **VB90 + n -1** – contains parameter n

The PLC routine designed to handle the serial port initializes the port in the first SCAN cycle, entering after that into the listen state. When a message is received, the data is transferred to the already mentioned memory zone and a copy is sent back to the calling PC.

For example, the PLC used with *Nicola* (*Siemens S7-200*) has 10 digital outputs in the basic module, labeled from Q0.0 to Q0.7 (output block 0), and from Q1.0 to Q1.1 (output block 1). To access those digital outputs, the command must specify the type of access (write or a read access), the signal number, and the signal value in the case of a write access (check Table 1.1).

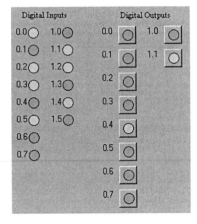

```
'Activates/deactivates digital outputs from block 0
Private Sub q0_Click(Index As Integer)
  If fq0(Index) = False Then
    com.Output = Chr(200)+Chr(120+Index)+Chr(10)
    fq0(Index) = True
  Else
    com.Output = Chr(201)+Chr(120+Index)+Chr(10)
    fq0(Index) = False
  End If
End Sub

' Activates/deactivates digital outputs from block 1
Private Sub q1_Click(Index As Integer)
  If fq1(Index) = False Then
    com.Output = Chr(200)+Chr(130+Index)+Chr(10)
    fq1(Index) = True
  Else
    com.Output = Chr(201)+Chr(130+Index)+Chr(10)
    fq1(Index) = False
  End If
End Sub
```

The serial port interrupt service routine stores the messages received from the PLC in the variables:
bq00 – digital output signals of block 0
bq10 – digital output signals of block 0
bi00 – digital input signals of block 0
bi10 – digital input signals of block 0

The routine **rio_click** represents the received information at the user panel using colors: yellow (activated), gray (deactivated).

Digital Outputs

Digital Inputs

When a message arrives, the service routine calls "rio_click" to present the information:

```
Private Sub com_OnComm()
  get_com_message
  rio_Click
End Sub
```

```
' Shows IO state
Private Sub rio_Click()
  Dim i As Integer
  For i = 0 To 7
    If (bq00 And 2 ^ i) = 2 ^ i Then
      q0(i).Picture = Ion
      fq0(i) = True
    Else
      q0(i).Picture = Ioff
      fq0(i) = False
    End If
  Next i
  For i = 0 To 1
    If (bq10 And 2 ^ i) = 2 ^ i Then
      q1(i).Picture = Ion
      fq1(i) = True
    Else
      q1(i).Picture = Ioff
      fq1(i) = False
    End If
  Next i
  For i = 0 To 7
    If (bi00 And 2 ^ i) = 2 ^ i Then
      i0(i).Picture = Ion
    Else
      i0(i).Picture = Ioff
    End If
  Next i
  For i = 0 To 5
    If (bi10 And 2 ^ i) = 2 ^ i Then
      i1(i).Picture = Ion
    Else
      i1(i).Picture = Ioff
    End If
  Next i
End Sub
```

Figure 1.12 PC software designed to access IO signals

Consequently, to change the state of Q1.1 to 1 the following command should be sent (Table 1.1):

200 131 255 255 10

where *"200"* specifies a digital write access, *"131"* specifies the output Q1.1, *"255"* is a null command/parameter and "10" is the end-of-message character. The software for this example, including the both the PLC and the PC side, is presented in Figures 1.12 and 1.13 (the PC part was coded using *Microsoft Visual Basic .NET2003*, and the PLC part was coded using the *Siemens PLC S7-200* programming tool called *Microwin 3.2*).

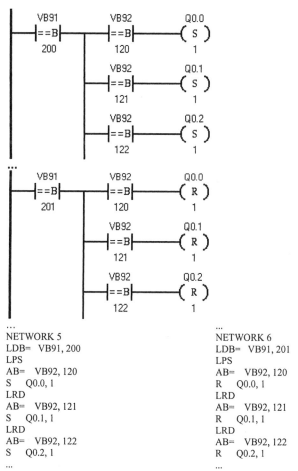

```
...                                        ...
NETWORK 5                                  NETWORK 6
LDB= VB91, 200                             LDB= VB91, 201
LPS                                        LPS
AB= VB92, 120                              AB= VB92, 120
S    Q0.0, 1                               R    Q0.0, 1
LRD                                        LRD
AB= VB92, 121                              AB= VB92, 121
S    Q0.1, 1                               R    Q0.1, 1
LRD                                        LRD
AB= VB92, 122                              AB= VB92, 122
S    Q0.2, 1                               R    Q0.2, 1
...                                        ...
```

Figure 1.13 PLC code to activate/deactivate digital outputs. Due to space limitations, only the code for the first three outputs of the digital block 0 is presented.

1.3.2.2 Software for the On-board PC

The software for the on-board PC was designed to control the robot, and to interface with the remote user connected to the robot's on-board computer using a wireless network connection (Figure 1.14).

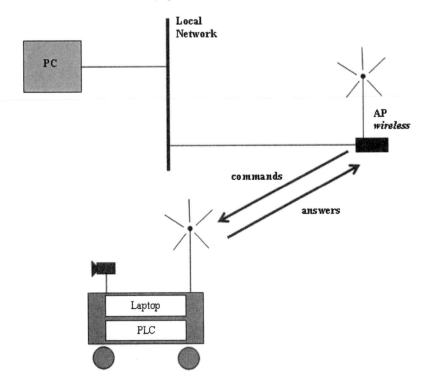

Figure 1.14 Overview of the system used to operate the robot *Nicola*

The on-board user interface software is a TCP/IP socket server that listens on a specific port, accepts and validates user connections, and processes the commands sent by the remote client. Those commands have the following basic syntax:

rx command parameter_1 parameter_2 ... parameter_n

where, *rx* specifies the robot (for example, r1), *command* is a string that specifies the command to be executed (Table 1.2), and *parameter_i* is the set of parameters associated with the particular command.

Figure 1.15 shows the shell of the TCP/IP server developed for the on-board computer. The panel functions enable the user to quickly access the local robot functions, and the TCP/IP server included in the application implements the interface for remote users.

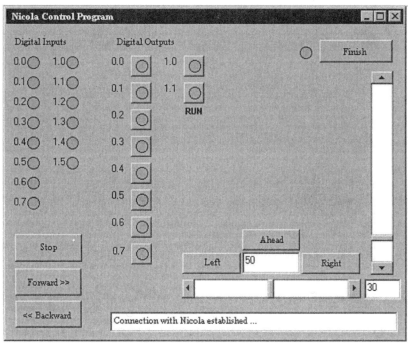

Figure 1.15 TCP/IP server used to operate the robot *Nicola*: listens to connections on port 54321, validates connections, and process commands

Table 1.2 List of commands available from the on-board TCP/IP server

Command	Parameter	Description
INFO	-	Supervision message.
VELC	Valor (0-255)	Commands the robot velocity: 0 (min.) to 255 (máx.).
STOP	-	Stop command.
AVAN	-	Commands the motors to move in the positive (forward) direction.
RECU	-	Commands the motors to move in the negative (backward) direction.
FRNT	-	Commands the motors to move straight ahead/backward, i.e., clears any steering direction.
DIRT	-	Turns right at 50%, i.e., the actual velocity of the left motor is kept and the velocity of the right motor is reduced by 50%.
ESQD	-	Turns left at 50%, i.e., the actual velocity of the right motor is kept and the velocity of the left motor is reduced by 50%.
DIRD	Valor (0-100)	Turns right by the specified amount.
DIRE	Valor (0-100)	Turns left by the specified amount.

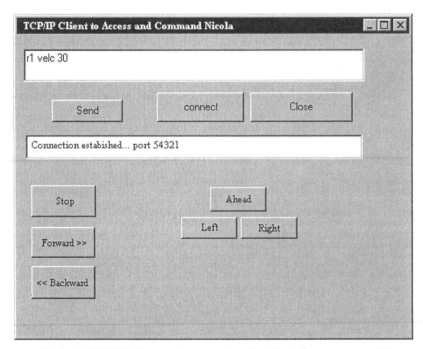

Figure 1.16 TCP/IP client used to operate the robot *Nicola* from any remote PC

Figure 1.16 shows a simple TCP/IP client example that can be used to command remotely the robot *Nicola*. This example offers to the user the possibility to execute simple commands like start and stop, move forward or backward, turn left and right (with a specified steering angle), or move straight ahead and regulate the robot's speed.

1.3.2.3 Feedback from the On-board Webcam

The robot *Nicola* is equipped with a *webcam* to register images of it's operation and to help the remote user command it in situations were the robot is not in sight. It's very easy to get images and video streams from a *webcam* and there are a lot of software packages and tools to do that. Here the *Microsoft Visual SDK 1.2* is used because it is an open source *SDK*, and because it integrates well with the development environment used to write the software: the *Microsoft Visual Studio .NET2003*.

Since the video feed is installed on the robot, there's also the problem of sending the obtained images from the on-board computer to the remote computer, using the data rate more adjusted to the capacity of the wireless link.

Again we opted to build a TCP/IP server to work as the image service. Basically this server is able to capture images and save those images in the hard disk of the on-board computer. These files can then be shared with the remote computer using an FTP connection or simply by sharing the directory. Using a mechanism like a semaphore it is possible to avoid having the two computers accessing the file at the same time, i.e., by the on-board computer that generates the file and by the remote computer that reads the file and presents it to the user. The image refresh rate depends on the communication speed and availability, but also on the size of the image. Nevertheless, it is possible to have rates up to 10 frames per second. Live streams, of about 30 to 40 frames per second, are only possible for the on-board computer since it was decided to avoid sending streams over the TCP/IP connection. This was a decision for simplicity, but also a practical decision: Live streams are really not necessary for this application.

The TCP/IP image server implements the following basic services:

1. Specify the vision provider, namely the driver that will be used to capture the image. In this example the Webcam uses a *Video for Windows* (VFW) driver
2. Start/stop the acquisition service
3. Obtain the actual image and save it to the on-board hard disk

The image server (Figure 1.17) listens at the port 54322 for messages starting with the character "@" and ending with the character "#". For example, the command message to obtain the actual image is:

$$@IMAGE\ rita\ beatriz\ dina\#$$

where, *IMAGE* is the command, *rita* is the username, *beatriz* is the password and *dina* is the name of the file where the image should be saved. The TCP/IP client will present the image only if the answer from the server matches exactly the command sent. Any other situation is considered an error.

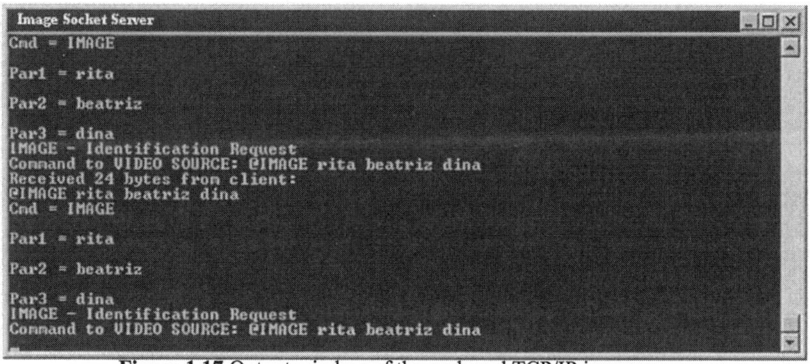

Figure 1.17 Output window of the on-board TCP/IP image server

Basically, the TCP/IP image client (Figure 1.18) has one button for each available service and shows the obtained image and the refresh rate. The method used to avoid simultaneous access to the image file between the two computers was a 50ms timer. The timer interrupt service routine performs alternatively the call to acquire the image and the call to get the file from the on-board computer, avoiding the simultaneous access to the image file. This means that a new image is obtained every 100 ms. Consequently, the only limitation to the refresh rate is the throughput of the communication link.

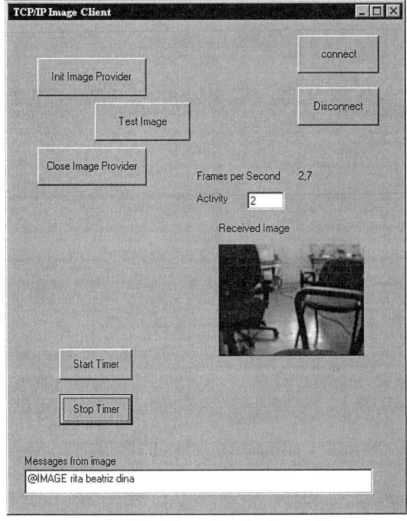

Figure 1.18 TCP/IP image client used on the remote PC

This simple example, which explores industrial equipment to build a useful mobile robot, shows clearly that robotics is a very interesting subject to learn and play with science and technology. It also shows that the concepts are accessible and can be explored by any student or engineer. The main objective of this section was to motivate readers to explore this book, because it'll show how things work and can be implemented in a practical way, with enough detail for those readers who want to explore further.

1.4 Using Robotics to Work

The industrial robotic system presented in this section was designed to execute the task of removing the excess of PVC material from automobile glasses, which accumulates during the glass manufacturing cycle. In fact, most of the automobile glasses, namely front, rear, and roof glasses, are composed of two sheets of glass joined by a layer of PVC. For proper assembly, and to ensure proper joining of the PVC to the glass while maintaining transparency, the glass goes through a heating process, followed by a considerable period inside a pressure chamber. This process generates a very stiff excess of PVC on the borders of the glass that must be carefully removed because it alters the dimensions of the glass, causing difficulties in assembling it in the car body, not to mention the aesthetic implications.

Figure 1.19 Robotic glass deburring system

Traditionally, this excess of PVC is removed by hand using small cutting devices. Nevertheless, for highly efficient plants, this is not desirable since it slows down production, and requires very high concentration from operators so they don't touch and damage the glass with the cutting device. Consequently, the process is very risky for the quality of the final product. Furthermore, with recent car designs, some glasses are glued directly in the chassis without any exterior rubber, mainly with roof, front, and rear glasses. Here the requirements for perfect PVC removal are even higher, which demands an automatic procedure to execute it.

The system (Figure 1.19) designed to handle the operation described above is composed of [12]:

1. Two industrial robots ABB IRB6400 equipped with the S4C+ controllers
2. Specially designed electric-pneumatic grippers to hold firmly the glasses
3. Two automatic deburring belts controlled by the robot's controller IO system
4. One industrial PLC (*Siemens S7-300*) that manages the cell logic and the interface to the adjacent industrial systems, providing to the robot controllers the necessary state information and the interface to the factory facilities
5. One personal computer to command, control and monitor the cell operation

The system works as follows: The first robot verifies if conveyor 1 (Figure 1.19) is empty and loads it with a glass picked from the pallet in use. The system uses a rotating circular platform to hold three pallets of glasses, enabling operators to remove empty pallets and feed new ones without stopping production. After releasing the glass, the robot pre-positions to pick another glass, which it does when the conveyor is again empty. If the working glass model requires deburring, then the centering device existing in the conveyor is commanded to center the glass so that the second robot could pick up the glasses in the same position. With the glass firmly grasped, the deburring robot takes it to the deburring belts and extracts the excess PVC by passing all the glass borders on the surface of the deburring belt. When the task is finished, the robot delivers the glass on conveyor 2, and proceeds to pick another glass.

The deburring velocity, pressure, trajectory, *etc.*, is stored in the robot system on a database sorted by the glass model, which makes it easy to handle several models. Programming a new model into the system is also very simple and executed by an authorized operator. There is a collection of routines that take the robot to pre-defined positions, adjusted by the given dimensions of the glass, allowing the operator to adjust and tune positions and trajectories. He can then "*play*" the complete definition and repeat the teaching procedure until the desired behavior is obtained. This means being able to control the robot's operation with the controller in automatic mode, which is obtained by including some teach-pendant features in the process for operator interface.

Another important feature of this robotic system is the ability to adjust production online, adapting to production variations. This objective is obtained by using a client-server architecture, which uses the cell computer (*client*) to parameterize the software running on the robot controller (*server*). That can be achieved by offering the following services from the robot server to the clients:

1. All planned system functionalities by means of general routines, callable from the remote client using variables that can be accessed remotely
2. Variable access services that can be used remotely to adjust and parameterize the operation of the robotic system

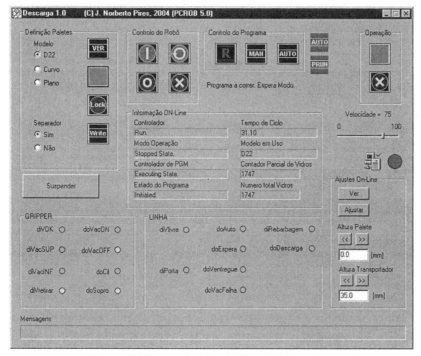

Figure 1.20 Operator interface for de-palletizing robot

With these features implemented and with a carefully designed operator interface (Figure 1.20 and Figure 1.21) and robot server software, it's possible to achieve a system that requires limited human intervention related with adjustment tasks to cope with production variations. Since a remote interface is used (Figures 1.20 and 1.21), the necessary adjustments are executed *online* without stopping production. Those operations include:

1. Adjusting the deburring angle, *i.e.*, the angle between the border of the glass and the deburring belt. The angle introduced is added to the programmed one, so that zero degrees means keeping the programmed angle unchanged

2. Adjusting the force on the belt during the deburring operation (adjusted by position). The commanded value is entered in millimeters and updates the actual position in the direction perpendicular to the belt and parallel to the surface of the glass
3. Adjusting the deburring speed
4. Maintenance procedures necessary to change the belts after the planned deburring cycles

The de-palletizing robot requires less parameterization because it executes a very simple operation. Other than that, the gripper adapts to the surface of every model of glass, using presence sensors strategically placed near two suction cups (see Figure 1.19), with the objective of having an efficient de-palletizing operation. Nevertheless, the operator is able to change the velocity of the process by stating a slow, fast, or very fast cycle to adjust to production needs, suspend and resume operations, adjust the way the robot approaches the surface of the glass, *etc.*. These adjustments are necessary to obtain the most efficient operation in accordance with the observed production conditions, to solve daily problems, and to cope with production variations.

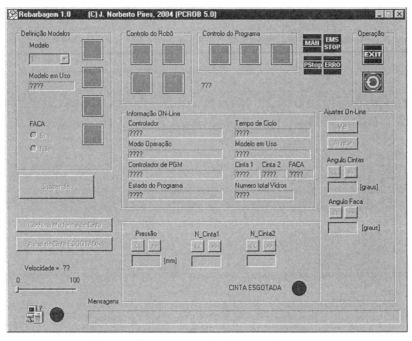

Figure 1.21 Operator interface for deburring robot

Finally, it is important to mention that the robot is equipped with a force/torque sensor mounted on the wrist. The objective is to adjust automatically the model setup introduced by the operator, correcting the points where the measured force between the belt and the glass exceeds the recommended values, attempting to avoid damage to the glass and to increase the deburring efficiency. This procedure is active during the process of applying a new model, and also during production, if explicitly activated by the operator, constituting an automatic correcting feature.

The system has worked for some time and proved to be very simple to operate, showing also quick adaptation from operators [12,18]. The adjusting features added to the system proved to be very helpful, allowing the company to respond in a timely fashion to production changes, avoiding variations in the quality of the final product, and to introduce quickly new models into the production database. Since the models are identified automatically, using barcode readers placed on the pallet platform, the system works continuously without operator intervention. The only thing needed is to feed the system with pallets full of glasses, removing the empty ones. That operation is done periodically with the help of electro-mechanical fork lift trucks.

Most of the features presented in this example will be explored in this book for robotic welding applications, namely the capacity to simulate the procedure, the capacity to adjust online and change parameterization, the capacity to monitor the system, and specify the sequence of operations, and so on.

This example shows clearly the advantages of using robots with actual manufacturing platforms and the importance of carefully designing the manufacturing systems, and integrating intelligent sensors, actuators, and the human factor. This final aspect related with HMI (human-machine interface) is fundamental in any manufacturing system and somehow a measure of its success, since these systems need a very efficient way to operate with humans in a way to expose system features and allow the users to explore the system capabilities to the maximum extent [12-18].

1.4.1 Using an Offline Simulation Environment

Using offline programming and simulation environments may be useful to develop and especially to optimize industrial manufacturing systems. Frequently the system is not available for online utilization, which calls for the possibility to work with graphical models of the manufacturing cell under study. The industrial deburring system presented in this section (Figures 1.22 and 1.23) was optimized using a graphical offline tool (*RobotStudio 5* from *ABB Robotics*), although the 3D drawings of several components of the cell were designed using *SolidWorks*.

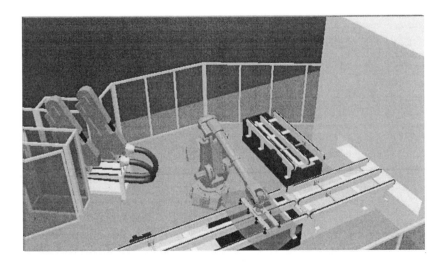

Figure 1.23 Analyzing the glass deburring process on the graphical environment

The utilization of offline packages has some advantages:

- If carefully designed, the graphical model constitutes a powerful tool to continuously develop the system without stopping production

- It allows the system engineer to simulate and optimize the solutions before testing them on the real cell for final implementation
- It constitutes a powerful tool to analyze new production scenarios, with new products, new production sequences, etc., before testing them or even before proposing them to the production team
- It constitutes a nice environment to demonstrate to customers the viability of certain type of production, cycle time, *etc*
- Since this type of environment runs a virtual robot controller, it allows the user to develop software and try it on the graphical model

The only disadvantage is the correlation between the graphical model and the real system. This means that the system engineer needs to carefully calibrate the system using precise data from the cell. This will allow him to export code directly to the cell and have it working with only minor calibration and routine checking.

1.5 Statistics of Robotic Workers

There are at least 800 000 robots working in industry worldwide (Table 1.3), but since statistics are very difficult to obtain in several countries, the real number should be over 1 million units operating all over the world [23]. Considering the statistics from 2003 [23], the lead country pushing its economy using robots is Japan, with around 350 000 robots operating, followed by the European Union, with around 250 000 robots in action, and the United States with around 112 000 robots. In Europe, Germany is the lead country with 112 700 units operating (matching the United States), followed by Italy (50 000 robots), France (26 000 robots) and Spain (20 000 robots).

Table 1.3 Robot operational stock at the end of the year (2001-2003) with a forecast for the period 2004-2007

Country	Operational Stock at the End of the Year			
	2002	2003	2004	2007
Japan	350 169	348 734	352 200	349 400
USA	103 515	112 390	121 300	145 100
EU	233 769	249 200	266 100	325 900
Germany	105 212	112 693	121 500	151 400
Italy	46 881	50 043	53 100	151 400
France	24 277	26 137	28 400	35 900
Spain	18 352	19 847		
Portugal	1 282	1 367		

Source: IFR – International Federation of Robotics [23]

In 1990, the installation of new industrial robots in the European Union was only 20% of the new installations reported from Japan. The USA had only 7% of new installations when compared with Japan. Nevertheless, this gap was reduced significantly and currently both EU and USA grow at approximate rates when compared with Japan, being sometimes higher than the Japanese rates. For example, in the period 2001-2002, the European Union installed more robots than Japan, but in 2003 the Japanese recovered the first place. This evolution of the European and North American robot installations reveals itself in the operational stock. The European stock evolved from 23% of that of Japan in 1990 to almost 72% in 2003. The figures for the USA show an evolution from 12% in 1990 to 32% in 2003, respectively.

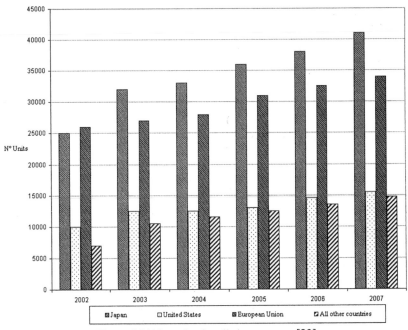

Figure 1.24 New robot installations per year [23]

The IFR forecast for 2007 expects a steady growth of robot installations in the European Union (6.1% per year) and in the United states (5.8% per year). Although Japan's new installations experienced different growth rates in the period 1999-2001, a significant recovery started in 2002 and a steady growth rate is expected at least until 2007 (5.7% per year).

Robots are becoming very common in any industrial installation (Figure 1.23 shows the number of robots per 10 000 workers for the motor vehicle industry, one of the most successful areas of robot operation) where they cooperate with human workers to achieve better efficiency and productivity. The pressure to invest in robots, namely regarding cost savings, increases in productivity and quality, and transferring dangerous tasks from humans to machines, i.e., to remain competitive in the global market, configures a scenario where humans and robots share the working space. In fact, in the beginning of the 21st century, robots are already human coworkers and successful installations must consider carefully the human-robot interaction and handle it as efficiently as possible.

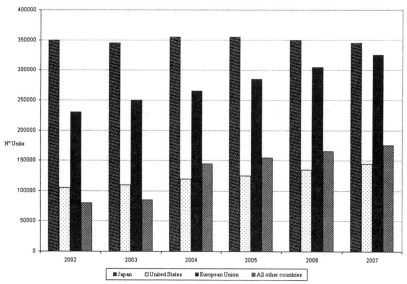

Figure 1.25 Operational stocks at the end of the year [23]

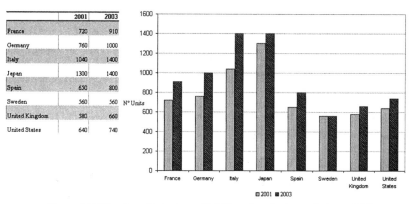

Figure 1.26 Number of robots per 10 000 workers in the car industry [23]

Consequently, industrial robots fit well with the two main challenges faced currently by modern manufacturing: more quality at lower prices and the need to improve productivity. Those are the requirements to keep manufacturing plants in developed countries, rather in the low-salary regions of the world. Other very important characteristics of manufacturing systems are flexibility and agility since companies need to respond to a very dynamic market with products that have low life-cycles due to fashion tendencies and worldwide competition.

So, manufacturing companies need to respond to market needs efficiently, keeping their products competitive. This requires a very efficient and controlled manufacturing process, where focus is on automation, computers and software.

The final objective is to achieve semi-autonomous systems, i.e., highly automated systems that require only minor operator intervention. In many industries, production is closed tracked in any part of the manufacturing cycle, which is composed by several in-line manufacturing systems that perform the necessary operations to transform the raw materials into a final product. In many cases, if properly designed, those individual manufacturing systems require simple parameterization to execute the tasks they are designed to execute. If that parameterization can be commanded remotely by automatic means from where it is available, then the system becomes almost autonomous in that operator intervention is reduced to the minimum and essentially needed for error and maintenance situations. Human and machines can cooperate doing their own tasks, more or less autonomously, and interface more closely when required by the manufacturing process.

A system like this will improve efficiency and agility, since it is less dependent on human operators. Also, since those systems are built under distributed frameworks, based on client-server software architectures that require a collection of functions that implement the system functionality, it is easier to change production by adjusting parameterization (a software task now) which also contributes to agility. Furthermore, since all information about each item produced is available in the manufacturing tracking software, it is logical to use it to command some of the shop floor manufacturing systems, namely the ones that require simple parameterization to work properly. This procedure would take advantage of the available information and computing infrastructure, avoiding unnecessary operator interfaces to command the system. Also, further potential gains in terms of flexibility and productivity are evident.

1.6 Overview of the rest of the book

This book is about industrial robot programming in the beginning of twentieth first century. It focuses on the important aspects of designing and building robotic manufacturing cells, which explore the capabilities of the actual industrial equipment, and the available computer and software technologies. Special attention will be paid to exploring the available input devices and systems that can be used to create more efficient human-machine interfaces, namely to the programming, control, and supervision tasks performed by non-technical personnel.

Chapter Two ("Robot Manipulators and Control Systems") introduces most of the industrial robotic equipment currently available, namely aspects related with industrial robotic manipulators, their control systems and programming environments. In the process, two specific manipulators will be considered closely since both will be used in many examples presented in the rest of the book.

Chapter Three ("Software Interfaces") discusses software interfaces that can be used to develop distributed industrial manufacturing cells. It covers the mechanisms and techniques used to interface robots with computers, as well as intelligent sensors, actuators, other factory resources, production management software, and so on. The software discussed in this chapter is used in all the examples presented in the book, and is the core of several industrial and laboratory applications.

Chapter Four ("Interface Devices and Systems") presents an overview of several available devices and systems that can be used to program, control, and supervise industrial robotic manufacturing cells. The intention here is to show that these interfaces and systems are available and to demonstrate, with application examples, how they can be explored to design solutions easier to use and program by non-technical operators.

Chapter Five ("Industrial Manufacturing Systems") is dedicated to a few application examples designed and implemented recently by the author of this book. The applications are described in detail to enable the interested reader to explore further. Although the selected examples were designed for specific applications, and carefully tuned for the industry in which they are currently used, the discussion is kept general since most of the problems addressed are common to many industries.

Finally, **chapter six** ("Final Notes") presents a brief summary of the concepts and ideas presented in this book, and lists a few possible actions that the interested reader can follow to learn more about this important area of modern engineering.

A good collection of references is also presented at the end of each chapter to enable the reader to explore further.

1.7 References

[1]   Pires, JN, "Welding Robots. Technology, systems issues and applications", Springer, 2005.
[2]   Kusiak, A, "Computational Intelligence in Design and Manufacturing", John Wiley & Sons, 2000.
[3]   Halsall F., "Data Communications, Computer Networks and Open Systems", Third Edition, Addison-Wesley, 1992.
[4]   Tesla, N, "My Inventions: Autobiography of Nicola Tesla", Willinston, VT: Hart Brothers, 1983.
[5]   Rosheim, M, "Robot Evolution: The Development of Anthrobots", New York: John Willey & Sons, 1994.

[6]   Rosheim, M, "In the Footsteps of Leonardo", IEEE Robotics and Automation Magazine, June 1997.

[7]   Pedretti, C, "Leonardo Architect", Rizzoli International Publications, New York, 1981.

[8]   Mars Exloration WebSite (NASA), http://mars.jpl.nasa.gov

[9]   Mclennan Ltd., Precision Motion Control, http://www.mclennan.co.uk/

[10]   Siemens, Micro Automation SIMATIC S7-200, www.siemens.com/s7-200

[11]   Robot Nicola WebSite, http://robotics.dem.uc.pt/norberto/nova/nicola.htm

[12]   Pires, JN, "Semi-autonomous Manufacturing Systems: the role of the HMI software and of the manufacturing tracking software", IFAC Journal on Mechatronics, accepted for publication on Vol. 15, to appear in 2005.

[13]   Pires, JN, Sá da Costa JMG, "Object Oriented and Distributed Approach for Programming Robotic Manufacturing Cells", IFAC Journal on Robotics and Computer Integrated Manufacturing, February 2000.

[14]   Pires, JN, Paulo, S, "High-efficient de-palletizing system for the non-flat ceramic industry", Proceedings of the 2003 IEEE International Conference on Robotics and Automation, Taipei, 2003.

[15]   Pires, JN, "Object-oriented and distributed programming of robotic and automation equipment", Industrial Robot, An International Journal, MCB University Press, July 2000.

[16]   Pires, JN, "Interfacing Robotic and Automation Equipment with Matlab", IEEE Robotics and Automation Magazine, September 2000.

[17]   Pires, JN, "Force/torque sensing applied to industrial robotic deburring", Sensor Review Journal, MCB University Press, July 2002.

[18]   Pires, JN, Godinho, T, Ferreira, P, "CAD interface for automatic robot welding programming", Sensor Review Journal, MCB University Press, July 2002.

[19]   Bloomer, J, "Power Programming with RPC", O'Reilly & Associates, Inc., 1992.

[20]   Box, D, "Essential COM", Addison-Wesley, 1998

[21]   Rogerson, D, "Inside COM", Microsoft Press, 1997.

[22]   Visual C++ .NET 2003 Programmers Reference, Microsoft, 2003 (reference can be found at Microsoft's Web site in the Visual C++ .NET location)

[23]   "World Robotics 2004 – Statistics, Market Analysis, Forecasts, Case Studies and Profitability of Robot Investment", International Federation of Robotics and the United Nations, 2004.

2

Robot Manipulators and Control Systems

2.1 Introduction

This book focuses on industrial robotic manipulators and on industrial manufacturing cells built using that type of robots. This chapter covers the current practical methodologies for kinematics and dynamics modeling and computations. The kinematics model represents the motion of the robot without considering the forces that cause the motion. The dynamics model establishes the relationships between the motion and the forces involved, taking into account the masses and moments of inertia, i.e., the dynamics model considers the masses and inertias involved and relates the forces with the observed motion, or instead calculates the forces necessary to produce the required motion. These topics are considered very important to study and efficient use of industrial robots.

Both the kinematics and dynamics models are used currently to design, simulate, and control industrial robots. The kinematics model is a prerequisite for the dynamics model and fundamental for practical aspects like motion planning, singularity and workspace analysis, and manufacturing cell graphical simulation. For example, the majority of the robot manufacturers and many independent software vendors offer graphical environments where users, namely developers and system integrators, can design and simulate their own manufacturing cell projects (Figure 2.1).

Kinematics and dynamics modeling is the subject of numerous publications and textbooks [1-4]. The objective here is to present the topics without prerequisites, covering the fundamentals. Consequently, a real industrial robot will be used as an example which makes the chapter more practical, and easier to read. Nevertheless, the reader is invited to seek further explanation in the following very good sources:

1. *Introduction to Robotics*, JJ Craig, John Willey and Sons, Chapters 2 to 7.

2. *Modeling and Control of Robotic Manipulators*, F. Sciavicco and B. Siciliano, Mcgraw Hill, Chapters 2 to 5.
3. *Handbook of Industrial Robotics*, 2nd edition, Shimon Nof, Chapter 6 written by A. Goldenberg and M. Emani.

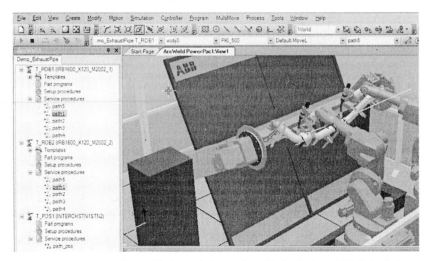

Figure 2.1 Aspect of a graphical simulation package (*RobotStudio* – ABB Robotics)

Another important practical aspect is the way how these topics are implemented and used by actual robot control systems. This chapter also reviews the fundamental aspects of robot control systems from the perspective of an engineer and of a system integrator. The objective is to introduce the main components and modules of modern robot control systems, by examining some of the control systems available commercially.

2.2 Kinematics

Actual industrial robot manipulators are very advanced machines exhibiting high precision and repeatability. It's common to have medium payload robots (16 to 20kg of payload) offering repeatability up to 0.1 mm, with smaller robots exhibiting even better performances (up to 0.01 mm). These industrial robots are basically composed by rigid links, connected in series by joints (normally six joints), having one end fixed (base) and another free to move and perform useful work when properly tooled (*end-effector*). As with the human arm, robot manipulators use the first three joints (arm) to position the structure and the remaining joints (wrist, composed of three joints in the case of the industrial manipulators) are used to orient the *end-effector*. There are five types of arms commonly used by actual industrial robot manipulators (Figure 2.2): *cartesian, cylindrical, polar, SCARA* and *revolution*.

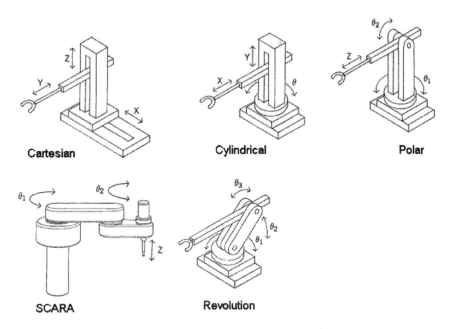

Figure 2.2 Types of arms used with actual robot manipulators

In terms of wrist designs, there are two main configurations (Figure 2.3):

1. pitch-yaw-roll (XYZ) like the human arm
2. roll-pitch-roll (ZYZ) or spherical wrist

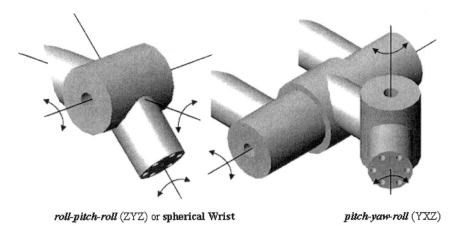

roll-pitch-roll (ZYZ) or **spherical Wrist**   *pitch-yaw-roll* (YXZ)

Figure 2.3 Wrist design configurations

The spherical wrist is the most popular because it is mechanically simpler to implement. Nevertheless, it exhibits singular configurations that can be identified and consequently avoided when operating with the robot. The trade between simplicity of robust solutions and the existence of singular configurations is favorable to the spherical wrist design, and that is the reason for its success.

The position and orientation of the robot's *end-effector* (tool) is not directly measured but instead computed using the individual joint position readings and the kinematics of the robot. Inverse kinematics is used to obtain the joint positions required for the desired *end-effector* position and orientation [1]. Those transformations involve three different representation spaces: a*ctuator space*, j*oint space* and c*artesian space*. The relationships between those spaces will be established here, with application to an ABB IRB1400 industrial robot (Figure 2.4). The discussion will be kept general for an anthropomorphic[1] manipulator with a spherical wrist[2].

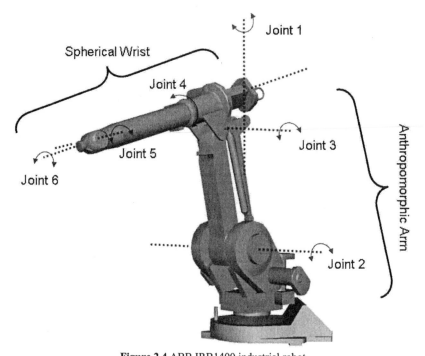

Figure 2.4 ABB IRB1400 industrial robot

---

[1] An anthropomorphic structure is a set of three revolute joints, with the first joint orthogonal to the other two which are parallel
[2] A spherical wrist has three revolute joints whose axes intersect at a single point

Table 2.1 *Denavit-Hartenberg* parameters for the IRB1400

Link	θ_i (°)	α_{i-1} (°)	a_{i-1} (mm)	d_i (mm)
1	θ_1 (0°)	0°	0	475
2	θ_2 (90°)	90°	150	0
3	θ_3 (0°)	0°	600	0
4	θ_4 (0°)	90°	120	720
5	θ_5 (0°)	-90°	0	0
6	θ_6 (0°)	90°	0	85 + d

where d is an extra length associated with the *end-effector*

Table 2.2 Workspace and maximum velocities for the IRB1400

Joint	Workspace (°)	Maximum Velocity (°/s)
1	+170° to -170°	110°/s
2	+70° to -70°	110°/s
3	+70° to -65°	110°/s
4	+150° to -150°	280°/s
5	+115° to -115°	280°/s
6	+300° to -300°	280°/s

Figure 2.5 represents, for simplicity, the robot manipulator axis lines and the assigned frames. The *Denavit-Hartenberg* parameters, the joint range and velocity limits are presented in Tables 2.1 and 2.2. The represented frames and associated parameters were found using Craig's convention [1].

2.2.1 Direct Kinematics

By simple inspection of Figure 2.5 it is easy to conclude that the last three axes form a set of *ZYZ Euler* angles [1,2] with respect to frame 4. In fact, the overall rotation produced by those axes is obtained from:

1. rotation about Z_4 by θ_4
2. rotation about $Y'_4=Z'_5$ by θ_5
3. rotation about $Z''_4=Z''_6$ by θ_6.[3]

which gives the following rotation matrix,

---

[3] Y'_4 corresponds to axis Y_4 after rotation about Z_4 by θ_4 and Z''_4 corresponds to Z_4 after rotation about $Y'_4=Z'_5$ by θ_5

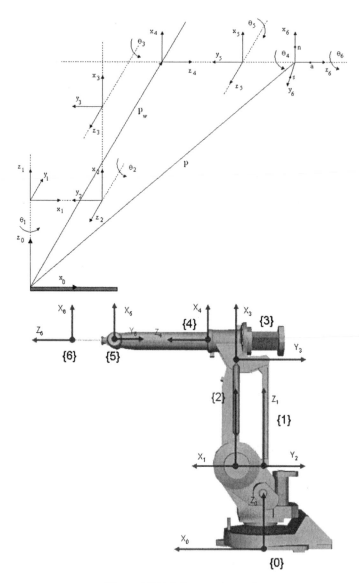

Figure 2.5 Link frame assignment

$$R_{Euler} = R_z(\theta_4).R_{y'4}(\theta_5).R_{z''4'}(\theta_6) =$$

$$= \begin{bmatrix} c_4 & -s_4 & 0 \\ s_4 & c_4 & 0 \\ 0 & 0 & 1 \end{bmatrix} \begin{bmatrix} c_5 & 0 & s_5 \\ 0 & 1 & 0 \\ -s_5 & 0 & c_5 \end{bmatrix} \begin{bmatrix} c_6 & -s_6 & 0 \\ s_6 & c_6 & 0 \\ 0 & 0 & 1 \end{bmatrix} \tag{2.1}$$

$$= \begin{bmatrix} c_4 c_5 c_6 - s_4 s_6 & -c_4 c_5 s_6 - s_4 c_6 & c_4 s_5 \\ s_4 c_5 c_6 + c_4 s_6 & -s_4 c_5 s_6 + c_4 c_6 & s_4 s_5 \\ -s_5 c_6 & s_5 s_6 & c_5 \end{bmatrix} = \begin{bmatrix} r_{11} & r_{12} & r_{13} \\ r_{21} & r_{22} & r_{23} \\ r_{31} & r_{32} & r_{33} \end{bmatrix} = R$$

The above rotation matrix R, in accordance with the assigned frame settings, should verify the following two equations:

$$R_6^3 = \begin{bmatrix} 1 & 0 & 0 \\ 0 & 0 & -1 \\ 0 & 1 & 0 \end{bmatrix}.R$$

$$R(\theta_4 = 0) = R_6^4 \tag{2.2}$$

The values of θ_4, θ_5 and θ_6 can be now obtained. Comparing r_{13} with r_{23} (considering $s_5 \neq 0$) results in,

$$\theta_4 = A \tan 2(r_{23}, r_{13}) \tag{2.3}$$

Squaring and summing r_{13} and r_{23} and comparing the result with r_{33} gives,

$$\theta_5 = A \tan 2(\sqrt{r_{13}^2 + r_{23}^2}, r_{33}) \tag{2.4}$$

if a positive square-root of $r_{13}^2 + r_{23}^2$ is chosen: this assumption limits the range of θ_5 to $[0, \pi]$.

Using the same argument now considering elements r_{31} and r_{32} the following is obtained for θ_6:

$$\theta_6 = A \tan 2(r_{32}, -r_{31}) \tag{2.5}$$

For $\theta_5 \in [-\pi, 0]$ the solution is:

$$\theta_4 = A \tan 2(-r_{23}, -r_{13})$$
$$\theta_5 = A \tan 2(-\sqrt{r_{13}^2 + r_{23}^2}, r_{33})$$
$$\theta_6 = A \tan 2(-r_{32}, r_{31}) \tag{2.6}$$

The IRB1400 is an anthropomorphic manipulator with spherical wrist. The anthropomorphic structure of the first three joints is the one that offers better dexterity to the robot manipulator. The first three joints are used to position the wrist. The orientation of the wrist is managed by the wrist spherical structure, which is also the one that gives higher dexterity. Using the link transformation matrix definition derived at [1],

$$
T_i^{i-1} = \begin{bmatrix}
c_i & -s_i & 0 & a_{i-1} \\
s_i c\alpha_{i-1} & c_i c\alpha_{i-1} & -s\alpha_{i-1} & -s\alpha_{i-1} d_i \\
s_i s\alpha_{i-1} & c_i s\alpha_{i-1} & c\alpha_{i-1} & c\alpha_{i-1} d_i \\
0 & 0 & 0 & 1
\end{bmatrix}
\tag{2.7}
$$

the direct kinematics of the ABB IRB1400 robot manipulator can be easily obtained (as presented in Figure 2.6).

$$
T_1^0 = \begin{bmatrix}
c_1 & -s_1 & 0 & 0 \\
s_1 & c_1 & 0 & 0 \\
0 & 0 & 1 & d_1 \\
0 & 0 & 0 & 1
\end{bmatrix}
\quad
T_2^1 = \begin{bmatrix}
-s_2 & -c_2 & 0 & a_1 \\
0 & 0 & -1 & 0 \\
c_2 & -s_2 & 0 & 0 \\
0 & 0 & 0 & 1
\end{bmatrix}
\quad
T_3^2 = \begin{bmatrix}
c_3 & -s_3 & 0 & a_2 \\
s_3 & c_3 & 0 & 0 \\
0 & 0 & 1 & 0 \\
0 & 0 & 0 & 1
\end{bmatrix}
$$

$$
T_4^3 = \begin{bmatrix}
c_4 & -s_4 & 0 & a_3 \\
0 & 0 & -1 & -d_4 \\
s_4 & c_4 & 0 & 0 \\
0 & 0 & 0 & 1
\end{bmatrix}
\quad
T_5^4 = \begin{bmatrix}
c_5 & -s_5 & 0 & 0 \\
0 & 0 & 1 & 0 \\
-s_5 & -c_5 & 0 & 0 \\
0 & 0 & 0 & 1
\end{bmatrix}
$$

$$
T_6^5 = \begin{bmatrix}
c_6 & -s_6 & 0 & 0 \\
0 & 0 & -1 & -d_6 \\
s_6 & c_6 & 0 & 0 \\
0 & 0 & 0 & 1
\end{bmatrix}
\quad
T_2^0 = \begin{bmatrix}
-c_1 s_2 & -c_1 c_2 & s_1 & a_1 c_1 \\
-s_1 s_2 & -s_1 c_2 & -c_1 & a_1 s_1 \\
c_2 & -s_2 & 0 & d_1 \\
0 & 0 & 0 & 1
\end{bmatrix}
$$

$$
T_3^0 = \begin{bmatrix}
-c_1 s_{23} & -c_1 c_{23} & s_1 & -a_2 c_1 s_2 + a_1 c_1 \\
-s_1 s_{23} & -s_1 c_{23} & -c_1 & -a_2 s_1 s_2 + a_1 s_1 \\
c_{23} & -s_{23} & 0 & a_2 c_2 + d_1 \\
0 & 0 & 0 & 1
\end{bmatrix}
$$

$$
T_4^0 = \begin{bmatrix}
-c_1 s_{23} c_4 + s_1 s_4 & c_1 s_{23} s_4 + s_1 c_4 & c_1 c_{23} & d_4 c_1 c_{23} - a_3 c_1 s_{23} - a_2 c_1 s_2 + a_1 c_1 \\
-s_1 s_{23} c_4 - c_1 s_4 & s_1 s_{23} s_4 - c_1 c_4 & s_1 c_{23} & d_4 s_1 c_{23} - a_3 s_1 s_{23} - a_2 s_1 s_2 + a_1 s_1 \\
c_{23} c_4 & c_{23} s_4 & s_{23} & d_4 s_{23} + a_3 c_{23} + a_2 c_2 + d_1 \\
0 & 0 & 0 & 1
\end{bmatrix}
$$

$$
T_6^3 = \begin{bmatrix}
c_4 c_5 c_6 - s_4 s_6 & -c_4 c_5 s_6 - s_4 c_6 & c_4 s_5 & d_6 c_4 s_5 + a_3 \\
s_5 c_6 & -s_5 s_6 & -c_5 & -d_6 c_5 - d_4 \\
s_4 c_5 c_6 + c_4 s_6 & -s_4 c_5 s_6 + c_4 c_6 & s_4 s_5 & d_6 s_4 s_5 \\
0 & 0 & 0 & 1
\end{bmatrix}
$$

$$T_6^4 = \begin{bmatrix} c_5c_6 & -c_5s_6 & s_5 & d_6s_5 \\ s_6 & c_6 & 0 & 0 \\ -s_5c_6 & s_5s_6 & c_5 & d_6c_5 \\ 0 & 0 & 0 & 1 \end{bmatrix} \text{ and } T_6^0 = \begin{bmatrix} r_{11} & r_{12} & r_{13} & p_x^0 \\ r_{21} & r_{22} & r_{23} & p_y^0 \\ r_{31} & r_{32} & r_{33} & p_z^0 \\ 0 & 0 & 0 & 1 \end{bmatrix} \text{ with,}$$

$r_{11} = ((s_1s_4 - c_1s_{23}c_4)c_5 - c_1c_{23}s_5)c_6 + (c_1s_{23}s_4 + s_1c_4)s_6$

$r_{12} = ((-s_1s_4 + c_1s_{23}c_4)c_5 + c_1c_{23}s_5)s_6 + (c_1s_{23}s_4 + s_1c_4)c_6$

$r_{13} = (-c_1s_{23}c_4 + s_1s_4)s_5 + c_1c_{23}c_5$

$r_{21} = ((-s_1s_{23}c_4 - c_1s_4)c_5 - s_1c_{23}s_5)c_6 + (s_1s_{23}s_4 - c_1c_4)s_6$

$r_{22} = ((s_1s_{23}c_4 + c_1s_4)c_5 + s_1c_{23}s_5)s_6 + (s_1s_{23}s_4 - c_1c_4)c_6$

$r_{23} = (-s_1s_{23}c_4 - c_1s_4)s_5 + s_1c_{23}c_5$

$r_{31} = (c_{23}c_4c_5 - s_{23}s_5)c_6 - c_{23}s_4s_6$

$r_{32} = (-c_{23}c_4c_5 + s_{23}s_5)s_6 - c_{23}s_4c_6$

$r_{33} = c_{23}c_4s_5 + s_{23}c_5$

$p_x^0 = ((-c_1s_{23}c_4 + s_1s_4)s_5 + c_1c_{23}c_5)d_6 + d_4c_1c_{23} - a_3c_1s_{23} - a_2c_1s_2 + a_1c_1$

$p_y^0 = ((-s_1s_{23}c_4 - c_1s_4)s_5 + s_1c_{23}c_5)d_6 + d_4s_1c_{23} - a_3s_1s_{23} - a_2s_1s_2 + a_1s_1$

$p_z^0 = d_6(c_{23}c_4s_5 + s_{23}c_5) + d_4s_{23} + a_3c_{23} + a_2c_2 + d_1$

Figure 2.6 Direct kinematics of an ABB IRB 1400 industrial robot

Having derived the direct kinematics of the IRB 1400, it's now possible to obtain the *end-effector* position and orientation from the individual joint angles $(\theta_1,\theta_2,\theta_3,\theta_4;\theta_5,\theta_6)$.

2.2.2 Inverse Kinematics

Inverse kinematics deals with the problem of finding the required joint angles to produce a certain desired position and orientation of the *end-effector*. Finding the inverse kinematics solution for a general manipulator can be a very tricky task. Generally they are non-linear equations. Close-form solutions may not be possible and multiple, infinity, or impossible solutions can arise. Nevertheless, special cases have a closed-form solution and can be solved.

The sufficient condition for solving a six-axis manipulator is that it must have three consecutive revolute axes that intersect at a common point: *Pieper* condition [5]. Three consecutive revolute parallel axes is a special case of the above condition, since parallel lines can be considered to intersect at infinity. The ABB IRB 1400 meets the *Pieper* condition due to the spherical wrist.

For these types of manipulators, i.e. manipulators that meet the *Pieper* condition, it is possible to decouple the inverse kinematics problem into two sub-problems: position and orientation. A simple strategy [1,2] can then be used to solve the inverse kinematics, by separating the position problem from the orientation problem. Consider Figure 2.5, where the position and orientation of the *end-*

effector is defined in terms of $\mathbf{p}$ and $R_6^0 = [n \ \ s \ \ a]$. The wrist position ($\mathbf{p_w}$) can be found using

$$\mathbf{p_w} = \mathbf{p} - d_6.\mathbf{a} \tag{2.8}$$

It is now possible to find the inverse kinematics for θ_1, θ_2 and θ_3 and solve the first inverse kinematics sub-problem, i.e, the position sub-problem. Considering Figure 2.7 it is easy to see that

$$\theta_1 = A \tan 2(p_{wy}, p_{wx})^4 \tag{2.9}$$

Once θ_1 is known the problem reduces to solving a planar structure. Looking to Figure 2.7 it is possible to successively write

$$p_{wx1} = \sqrt{p_{wx}^2 + p_{wy}^2} \tag{2.10}$$
$$p_{wz1} = p_{wz} - d_1 \tag{2.11}$$
$$p_{wx1'} = p_{wx1} - a_1 \tag{2.12}$$
$$p_{wy1'} = p_{wy1} \tag{2.13}$$
$$p_{wz1'} = p_{wz1} \tag{2.14}$$

and

$$p_{wx1'} = -a_2 s_2 + a_x c_{23'} \tag{2.15}$$
$$p_{wz1'} = a_2 c_2 + a_x s_{23'} \tag{2.16}$$

---

[4] Another possibility would be $\theta_1 = \pi + A \tan 2(p_{wy}, p_{wx})$ if we set $\theta_2 -> \pi - \theta_2$

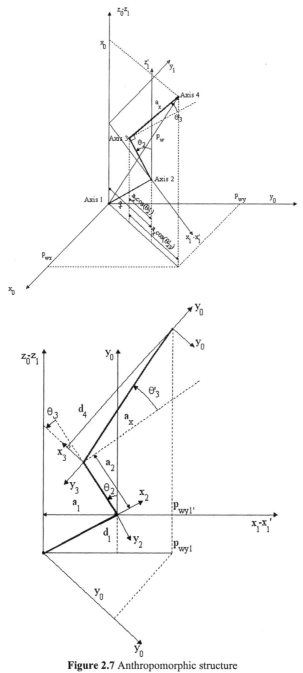

Figure 2.7 Anthropomorphic structure

Squaring and summing equations (2.15) and (2.16) results in

$$p_{wx1'}^2 + p_{wzl'}^2 = a_2^2 + a_x^2 + a_2 a_x s_{3'}$$ (2.17)

which gives

$$s_{3'} = \frac{p_{wxl'}^2 + p_{wzl'}^2 - a_2^2 - a_x^2}{2a_2 a_x}$$ (2.18)

Setting $c_{3'} = \pm\sqrt{1 - s_{3'}^2}$ the solution for θ'_3 will be

$$\theta'_3 = A \tan 2(s_{3'}, c_{3'})$$
$$\theta_3 = \theta'_3 - A \tan(a_3 / d_4)$$ (2.19)

Now, using θ'_3 in (2.15)-(2.16) results in a system with two equations with s_2 and c_2 unknowns:

$$p_{wxl'} = a_2 c_2 + a_x (c_2 c_{3'} - s_2 s_{3'})$$
$$p_{wzl'} = a_2 s_2 + a_x (s_2 c_{3'} + s_{3'} c_2)$$ (2.20)

Solving for s_2 and c_2 gives

$$s_2 = \frac{-(a_2 + a_x s_{3'}) p_{wxl'} + a_x c_{3'} p_{wzl'}}{a_2^2 + a_x^2 + 2a_2 a_x s_{3'}}$$ (2.21)

$$c_2 = \frac{(a_2 + a_x s_{3'}) p_{wzl'} + a_x c_{3'} p_{wxl'}}{a_2^2 + a_x^2 + 2a_2 a_x s_{3'}}$$ (2.22)

and the solution for θ_2 will be

$$\theta_2 = A \tan 2(s_2, c_2)$$ (2.23)

To solve the second inverse kinematics sub-problem (orientation), i.e., to find the required joint angles θ_4, θ_5 and θ_6 corresponding to a given *end-effector* orientation R_6^3, we simply take advantage of the special configuration of the last three joints. Because the orientation of the *end-effector* is defined by R_6^0, it's simple to get R_6^3 from,

$$R_6^3 = (R_3^0)^{-1}.R_6^0 = (R_3^0)^T.R_6^0$$ (2.24)

which gives

$$R_6^3 = \begin{bmatrix} -c_1 s_{23} & -s_1 s_{23} & c_{23} \\ -c_1 c_{32} & -s_1 c_{23} & -s_{23} \\ s_1 & -c_1 & 0 \end{bmatrix} \begin{bmatrix} a_{11} & a_{12} & a_{13} \\ a_{21} & a_{22} & a_{23} \\ a_{31} & a_{32} & a_{33} \end{bmatrix} = \begin{bmatrix} r_{11} & r_{12} & r_{13} \\ r_{21} & r_{22} & r_{23} \\ r_{31} & r_{32} & r_{33} \end{bmatrix} \quad (2.25)$$

with

$r_{11} = -c_1 s_{23} a_{11} - s_1 s_{23} a_{21} + c_{23} a_{31}$ $r_{12} = -c_1 s_{23} a_{12} - s_1 s_{23} a_{22} + c_{23} a_{32}$

$r_{13} = -c_1 s_{23} a_{13} - s_1 s_{23} a_{23} + c_{23} a_{33}$ $r_{23} = -c_1 c_{23} a_{13} - s_1 c_{23} a_{23} - s_{23} a_{33}$

$r_{33} = s_1 a_{13} - c_1 a_{23}$

$r_{21} = -c_1 c_{23} a_{11} - s_1 c_{23} a_{21} - s_{23} a_{31}$ $r_{22} = -c_1 c_{23} a_{12} - s_1 c_{23} a_{22} - s_{23} a_{32}$

$r_{31} = s_1 a_{11} - c_1 a_{21}$ $r_{32} = s_1 a_{12} - c_1 a_{22}$

It is now possible to use the previous result for the ZYZ Euler angles to obtain the solutions for θ_4, θ_5 and θ_6 .

For $\theta_5 \in [0, \pi]$ the solution is

$$\theta_4 = A \tan 2(r_{33}, r_{13})$$
$$\theta_5 = A \tan 2(\sqrt{r_{13}^2 + r_{33}^2}, -r_{23})$$
$$\theta_6 = A \tan 2(-r_{22}, r_{21}) \quad (2.26)$$

For $\theta_5 \in [-\pi, 0]$ the solution is

$$\theta_4 = A \tan 2(-r_{33}, -r_{13})$$
$$\theta_5 = A \tan 2(-\sqrt{r_{13}^2 + r_{33}^2}, r_{23})$$
$$\theta_6 = A \tan 2(r_{22}, -r_{21}) \quad (2.27)$$

2.3 Jacobian

In this section, the equations necessary to compute the jacobian of the ABB IRB1400 industrial robot are presented and the jacobian is obtained. Nevertheless, the discussion will be kept general for an anthropomorphic robot manipulator. In the process, the equations that describe the linear and angular velocities, static forces, and moments of each of the manipulator links are also presented and the corresponding developments applied to the selected robot.

The jacobian of any robot manipulator structure is a matrix that relates the *end-effector* linear and angular Cartesian velocities with the individual joint velocities:

$$V = \begin{bmatrix} v \\ w \end{bmatrix} = J(\theta).\dot{\theta} \tag{2.28}$$

where $J(\theta)$ is the jacobian matrix of the robot manipulator, $\dot{\theta} = \left[\dot{\theta}_1, \dot{\theta}_2, ..., \dot{\theta}_n\right]^T$ is the joint velocity vector, $v = [v_1, v_2, v_3]^T$ is the *end-effector* linear velocity vector, and $w = [w_1, w_2, w_3]^T$ is the *end-effector* angular velocity vector.

The jacobian is an n×m matrix, where n is the number of degrees of freedom of the robot manipulator and m is the number of joints. Considering an anthropomorphic robot manipulator with a spherical wrist, the corresponding jacobian will be a 6×6 matrix. Basically there are two ways to compute the jacobian:

1. By direct differentiation of the direct kinematics function with respect to the joint variables. This usually leads to the so-called *analytical jacobian*,

$$\dot{x} = \begin{bmatrix} \dot{p} \\ \dot{\phi} \end{bmatrix} = J_A(\theta).\dot{\theta} \tag{2.29}$$

where $\dot{p}$ is the time derivative of the position of the *end-effector* frame with respect to the base frame, $\dot{\phi}$ is the time derivative of the orientation vector expressed in terms of three variables (for instance, ZYZ Euler angles). Obviously, $\dot{p}$ is the translational velocity of the *end-effector* and $\dot{\phi}$ is the rotational velocity.

2. By computing the contributions of each joint velocity to the components of the *end-effector* Cartesian linear and angular velocities. This procedure leads to the *geometric jacobian*.

Generally, the analytical and geometrical jacobian are different from each other. Nevertheless, it is always possible to write

$$w = T(\phi).\dot{\phi} \tag{2.30}$$

where T is a transformation matrix from $\dot{\phi}$ to w. Once $T(\phi)$ is given, the analytical jacobian and geometric jacobian can be related by

$$V = \begin{bmatrix} I & 0 \\ 0 & T(\phi) \end{bmatrix}.\dot{x} = T_J(\phi).\dot{x} \tag{2.31}$$

which gives

$$J = T_J(\phi).J_A \tag{2.32}$$

Here the geometric jacobian will be calculated, because in the process the linear and angular velocities of each link will also be obtained. Nevertheless, the analytical jacobian should be used when the variables are defined in the operational space.

First the equations for the link linear and angular velocities and accelerations [1,2] will be obtained. Associating a frame to each rigid body, the rigid body motion can be described by the relative motion of the associated frames. Consider a frame {B} associated with a point D (Figure 2.8).

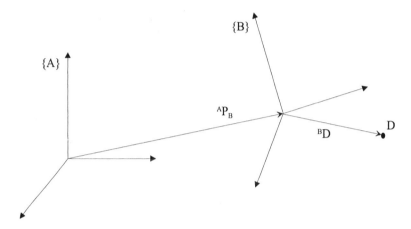

Figure 2.8 Describing point D relative to a stationary frame

The position vector of point D in frame {B} is ^BD and the relative velocity of D described about an arbitrary stationary frame {A} is [6],

$$^AV_D = {}^AV_B + {}^A_BR \ {}^BV_D \qquad (2.33)$$

If the relative motion between {A} and {B} is non-linear then (2.33) is not valid. The relative motion between two frames {A} and {B} has generally two components: a linear component ^AV_B and a non-linear component (the angular or rotational acceleration) $^A\Omega_B$ as in (Figure 2.9).

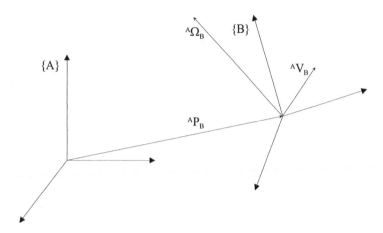

Figure 2.9 Relative motion between two frames {A} and {B}

In that general case it can be written [1,6,7],

$$^{A}V_D = {}^{A}V_B + {}^{A}_{B}R\ {}^{B}V_D + {}^{A}\Omega_B \times {}^{A}_{B}R\ {}^{B}D \tag{2.34}$$

where $^{A}V_D$ is the linear velocity of the origin of frame {B} about frame {A}, $^{A}_{B}R\ {}^{B}V_D$ is the linear velocity of point D about frame {B} expressed in terms of {A} (i.e., $^{A}_{B}R\ {}^{B}V_D = {}^{A}({}^{B}V_D)$), $^{A}\Omega_B \times {}^{A}_{B}R\ {}^{B}D = {}^{A}\Omega_B \times {}^{A}D$ is the linear velocity of point D about {A} expressed in terms of {A} as the result of the angular velocity $^{A}\Omega_B$ of {B} about {A}.

If D is stationary in {B} ($^{B}V_D = 0$) and the origins of {A} and {B} are coincident, i.e., the relative motion of D about {A} is only due to the rotation motion of {B} about {A} described by $^{A}\Omega_B$, then $^{A}V_D = {}^{A}\Omega_B \times {}^{A}_{B}R\ {}^{B}D$. This equation can also be obtained by differentiation of

$$^{A}D = {}^{A}_{B}R\ {}^{B}D \tag{2.35}$$

which yields

$$^{A}\dot{D} = {}^{A}_{B}\dot{R}\ {}^{B}D + {}^{A}_{B}R\ {}^{B}\dot{D} \tag{2.36}$$

or since in this special case $^{A}_{B}R\ {}^{B}\dot{D} = 0$,

$$^{A}V_D = {}^{A}_{B}\dot{R}\ {}^{B}D \tag{2.37}$$

Substituting in (2.37) $^B D = {}_B^A R^{-1}\ {}^A D$ results in

$$^A V_D = {}_B^A \dot{R}\ {}_B^A R^{-1}\ {}^A D \tag{2.38}$$

Because ${}_B^A R$ is an orthonormal matrix, we can write [1,7],

$${}_B^A \dot{R}\ {}_B^A R^{-1} = {}_B^A S \tag{2.39}$$

where ${}_B^A S$ is a *skew-symmetric* matrix associated with ${}_B^A R$.

Using (2.39) in (2.38) gives

$$^A V_D = {}_B^A S\ {}^A D \tag{2.40}$$

The skew-symmetric matrix ${}_B^A S$ defined in (2.39) is called *angular velocity matrix*.

Writing S as

$$S = \begin{bmatrix} 0 & -\Omega_z & \Omega_y \\ \Omega_z & 0 & -\Omega_x \\ -\Omega_y & \Omega_x & 0 \end{bmatrix} \tag{2.41}$$

and the vector Ω (3×1) as

$$\Omega = \begin{bmatrix} \Omega_x \\ \Omega_y \\ \Omega_z \end{bmatrix} \tag{2.42}$$

results in

$$S\,D = \begin{bmatrix} 0 & -\Omega_z & \Omega_y \\ \Omega_z & 0 & -\Omega_x \\ -\Omega_y & \Omega_x & 0 \end{bmatrix} . \begin{bmatrix} D_x \\ D_y \\ D_z \end{bmatrix} = \begin{bmatrix} -\Omega_z D_y + \Omega_y D_z \\ \Omega_z D_x - \Omega_x D_z \\ -\Omega_y D_x + \Omega_x D_y \end{bmatrix} = \Omega \times D \tag{2.43}$$

where $D = (D_x, D_y, D_z)^T$ is a position vector. The vector Ω associated with the angular velocity matrix is called an *angular velocity vector*. Using (2.43) and (2.40) gives

$$^A V_D = {}^A \Omega_B \times {}^A D \tag{2.44}$$

Considering now the linear and angular accelerations of each link, it's possible to write by direct differentiation of (2.34),

$$^A\dot{V}_D = {}^A\dot{V}_B + ({}^A_BR\ {}^BV_D)' + {}^A\dot{\Omega}_B \times {}^A_BR\ {}^BD + {}^A\Omega_B \times ({}^A_BR\ {}^BD)' \tag{2.45}$$

or since,

$$({}^A_BR\ {}^BV_D)' = {}^A_BR\ {}^B_{\ ,}\ {}_D + {}^A\Omega_B \times {}^A_BR\ {}^BV_D$$

and

$$({}^A_BR\ {}^BD)' = {}^A_BR\ {}^BV_D + {}^A\Omega_B \times {}^A_BR\ {}^BD,$$

$$^A\dot{V}_D = {}^A\dot{V}_B + {}^A_BR\ {}^B\dot{V}_D + 2\ {}^A\Omega_B \times {}^A_BR\ {}^BV_D +$$
$$+ {}^A\dot{\Omega}_B \times {}^A_BR\ {}^BD + {}^A\Omega_B \times ({}^A\Omega_B \times {}^A_BR\ {}^BD) \tag{2.46}$$

The above equation is the general equation for the linear acceleration of point D about {A} and expressed in terms of {A}. If ^BD is a constant vector (like in robotics applications) then equation (2.46) simplifies to

$$^A\dot{V}_D = {}^A\dot{V}_B + {}^A\dot{\Omega}_B \times {}^A_BR\ {}^BD + {}^A\Omega_B \times ({}^A\Omega_B \times {}^A_BR\ {}^BD) \tag{2.47}$$

because $^BV_D = {}^B\dot{V}_D = 0$.

If we consider a third frame {C}, with $^A\Omega_B$ being the angular velocity of {B} about {A} and $^B\Omega_C$ the angular velocity of {B} about {C}, then the angular velocity of {C} about {A} is,

$$^A\Omega_C = {}^A\Omega_B + {}^A_BR\ {}^B\Omega_C \tag{2.48}$$

Taking the derivative of (2.48) results in

$$^A\dot{\Omega}_C = {}^A\dot{\Omega}_B + ({}^A_BR\ {}^B\Omega_C)' = {}^A\dot{\Omega}_B + {}^A_BR\ {}^B\dot{\Omega}_C + {}^A\Omega_B \times {}^A_BR\ {}^B\Omega_C \tag{2.49}$$

This is a very useful equation to compute the angular acceleration propagation from link to link.

Let's apply this to a robot manipulator. As mentioned before we will consider only rigid manipulators with revolutionary joints, with the base frame as the reference frame.

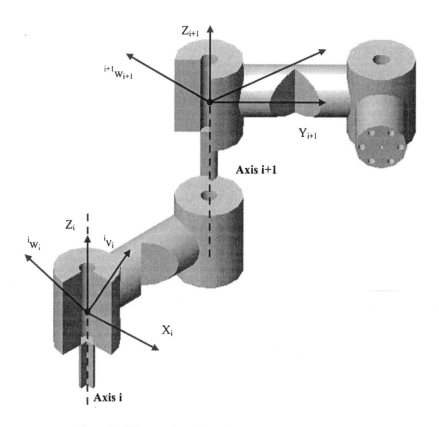

Figure 2.10 Linear and angular velocity vectors of adjacent links

The angular velocity of link (i+1), expressed in terms of {i}, is given by[5]

$$^{i}w_{i+1} = {}^{i}w_i + {}^{i}_{i+1}R \; \dot{\theta}_{i+1} \; {}^{i+1}Z_{i+1} \tag{2.50}$$

It is equal to the angular velocity of link (i) plus the angular velocity of joint (i+1) about Z_{i+1} expressed in terms of {i}.

Multiplying both sides of (2.50) by $^{i+1}_{i}R$ results in the angular velocity of link (i+1) expressed in terms of {i+1},

$$^{i+1}w_{i+1} = {}^{i+1}_{i}R \; {}^{i}w_{i+1} = {}^{i+1}_{i}R \; {}^{i}w_i + \dot{\theta}_{i+1} \; {}^{i+1}Z_{i+1} \tag{2.51}$$

---

[5] Note that wi+1 = 0Ωi+1 and that iwi+1 is the same quantity expressed in terms of {i}.

The linear velocity of the origin of {i+1}, expressed in terms of {i}, is given by

$$^{i}v_{i+1} = {}^{i}v_i + {}^{i}w_i \times {}^{i}P_{i+1} \tag{2.52}$$

It is equal to the linear velocity of the origin of {i} plus a term that results from the rotation of the link (i+1) about Z_{i+1}. The same solution can be obtained from (7) by making $^{i}P_{i+1}$ constant in {i}, i.e., by making $^{i}v_{i+1} = 0$.

Multiplying both sides of (2.52) by $^{i+1}_{i}R$ we get the linear velocity of link (i+1) expressed in terms of {i+1}

$$^{i+1}v_{i+1} = {}^{i+1}_{i}R \, ({}^{i}v_i + {}^{i}w_i \times {}^{i}P_{i+1}) \tag{2.53}$$

Applying (2.51) and (2.53) from link to link, the equations for $^{n}w_n$ and $^{n}v_n$ (where **n** is the number of joints) will be obtained. The equations for $^{0}w_n$ and $^{0}v_n$ can be obtained by pre-multiplying $^{n}w_n$ and $^{n}v_n$ by $^{0}_{n}R$:

$$^{0}w_n = {}^{0}_{n}R \; {}^{n}w_n \tag{2.54}$$

$$^{0}v_n = {}^{0}_{n}R \; {}^{n}v_n \tag{2.55}$$

It's also important to know how forces and moments distribute through the links and joints of the robot manipulator in a static situation, i.e., how to compute the forces and moments that keep the manipulator still in the various operating static configurations. Considering the manipulator at some configuration, the static equilibrium is obtained by proper balancing of the forces and moments applied to each joint and link, i.e., by cancelling the resultant of all the forces applied to the center of mass of each link (static equilibrium). The objective is to find the set of moments that should be applied to each joint to keep the manipulator in static equilibrium for some working configuration (Figure 2.11).

Considering,

f_i = force applied at link (i) by link (i-1)
n_i = moment in link (i) due to link (i-1)

the static equilibrium is obtained when

$$^{i}f_i - {}^{i}f_{i+1} = 0 \quad \text{and} \quad {}^{i}n_i - {}^{i}n_{i+1} - {}^{i}P_{i+1} \times {}^{i}f_{i+1} = 0 \tag{2.56}$$

i.e., when,

$$^{i}f_i = {}^{i}f_{i+1} \tag{2.57}$$

and

$$^{i}n_{i} = {}^{i}n_{i+1} + {}^{i}P_{i+1} \times {}^{i}f_{i+1} \tag{2.58}$$

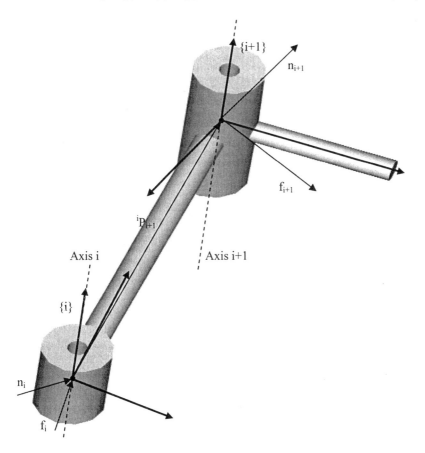

Figure 2.11 Static equilibrium: force balancing over link (i)

Writing the above equations in their own reference frame gives

$$^{i}f_{i} = {}^{i}_{i+1}R \; {}^{i+1}f_{i+1} \tag{2.59}$$

$$^{i}n_{i} = {}^{i}_{i+1}R \; {}^{i+1}n_{i+1} + {}^{i}P_{i+1} \times {}^{i}f_{i} \tag{2.60}$$

To compute the set of joint moments that hold the manipulator in static equilibrium we must obtain, for each joint (i), the projection of $^{i}n_{i}$ over the joint axis

$$\tau_{i} = {}^{i}n_{i}^{T} \; {}^{i}Z_{i} \tag{2.61}$$

Returning to the jacobian, from (2.54)-(2.55) it's possible to write

$$^0w_{i+1} = {}^0w_i + {}^i_{i+1}R \; (\dot{\theta}_{i+1} \; {}^{i+1}Z_{i+1}) \tag{2.62}$$

$$^0v_{i+1} = {}^0v_i + {}^0w_i \times {}^0P^i_{i+1} \tag{2.63}$$

Using (1) and (2.62)-(2.63) the ith column of the jacobian can be found to be

$$^0_nJ_i = \begin{bmatrix} z_i \times {}^0P^i_n \\ z_i \end{bmatrix} \tag{2.64}$$

Applying (2.62), (2.63), and (2.64) to the IRB1400 industrial robot, the equations presented in Figure 2.12 are obtained.

$$^0v_0 = 0 \quad ^0w_0 = 0 \quad ^0v_1 = 0 \quad ^0w_1 = \begin{bmatrix} 0 \\ 0 \\ \dot{\theta}_1 \end{bmatrix} \quad ^0v_2 = \begin{bmatrix} -a_1s_1\dot{\theta}_1 \\ a_1c_1\dot{\theta}_1 \\ 0 \end{bmatrix} \quad ^0w_2 = \begin{bmatrix} s_1\dot{\theta}_2 \\ -c_1\dot{\theta}_2 \\ \dot{\theta}_1 \end{bmatrix}$$

$$^0v_3 = \begin{bmatrix} (a_2s_1s_2 - a_1s_1)\dot{\theta}_1 - a_2c_1c_2\dot{\theta}_2 \\ (a_1c_1 - a_2c_1s_2)\dot{\theta}_1 - a_2c_2s_1\dot{\theta}_2 \\ -a_2s_2\dot{\theta}_2 \end{bmatrix} \quad ^0w_3 = \begin{bmatrix} s_1(\dot{\theta}_2 + \dot{\theta}_3) \\ -c_1(\dot{\theta}_2 + \dot{\theta}_3) \\ \dot{\theta}_1 \end{bmatrix}$$

$$^0v_4 =$$

$$\begin{bmatrix} (a_2s_2 - a_1 + a_3s_{23} - d_4c_{23})s_1\dot{\theta}_1 - (a_2c_2 + d_4s_{23} - a_3c_{23})c_1\dot{\theta}_2 - (d_4s_{23} + a_3c_{23})c_1\dot{\theta}_3 \\ (a_1 - a_2s_2 + d_4c_{23} - a_3s_{23})c_1\dot{\theta}_1 - (a_2c_2 + d_4s_{23} + a_3c_{23})s_1\dot{\theta}_2 - (d_4s_{23} + a_3c_{23})s_1\dot{\theta}_3 \\ (d_4c_{23} - a_3s_{23} - a_2s_2)\dot{\theta}_2 + (d_4c_{23} - a_3s_{23})\dot{\theta}_3 \end{bmatrix}$$

$$^0w_4 = \begin{bmatrix} s_1(\dot{\theta}_2 + \dot{\theta}_3) + c_1c_{23}\dot{\theta}_4 \\ -c_1(\dot{\theta}_2 + \dot{\theta}_3) + s_1c_{23}\dot{\theta}_4 \\ \dot{\theta}_1 + s_{23}\dot{\theta}_4 \end{bmatrix}$$

$$^0v_5 =$$

$$\begin{bmatrix} (a_2s_2 - a_1 + a_3s_{23} - d_4c_{23})s_1\dot{\theta}_1 - (a_2c_2 + d_4s_{23} - a_3c_{23})c_1\dot{\theta}_2 - (d_4s_{23} + a_3c_{23})c_1\dot{\theta}_3 \\ (a_1 - a_2s_2 + d_4c_{23} - a_3s_{23})c_1\dot{\theta}_1 - (a_2c_2 + d_4s_{23} + a_3c_{23})s_1\dot{\theta}_2 - (d_4s_{23} + a_3c_{23})s_1\dot{\theta}_3 \\ (d_4c_{23} - a_3s_{23} - a_2s_2)\dot{\theta}_2 + (d_4c_{23} - a_3s_{23})\dot{\theta}_3 \end{bmatrix}$$

$$^0w_5 = \begin{bmatrix} s_1(\dot{\theta}_2 + \dot{\theta}_3) + c_1c_{23}\dot{\theta}_4 + (c_1s_{23}s_4 + s_1c_4)\dot{\theta}_5 \\ -c_1(\dot{\theta}_2 + \dot{\theta}_3) + s_1c_{23}\dot{\theta}_4 + (s_1s_{23}s_4 - c_1c_4)\dot{\theta}_5 \\ \dot{\theta}_1 + s_{23}\dot{\theta}_4 - c_{23}s_4\dot{\theta}_5 \end{bmatrix} \quad ^0v_6 = \begin{bmatrix} ^0v_6(x) \\ ^0v_6(y) \\ ^0v_6(z) \end{bmatrix}$$

$$^0v_6(x) = ((a_2s_2 - a_1 + a_3s_{23} - d_4c_{23})s_1 + d_6((s_1s_{23}c_4 + c_1s_4)s_5 - s_1c_{23}c_5))\dot{\theta}_1 +$$
$$+ ((-a_2c_2 - d_4s_{23} - a_3c_{23}) - d_6(c_{23}c_4s_5 + s_{23}c_5))c_1\dot{\theta}_2 + (c_1(-d_4s_{23} - a_3c_{23}) - d_6(c_{23}c_4s_5 +$$
$$+ s_{23}c_5))\dot{\theta}_3 + d_6(s_1c_4s_5 + c_1s_4s_5s_{23})\dot{\theta}_4 + d_6(s_1c_5s_4 - c_1c_{23}s_5 - c_1c_4c_5s_{23})\dot{\theta}_5$$

$${}^0v_6(y) = ((a_1 - a_2s_2 + d_4c_{23} - a_3s_{23})c_1 + ((-c1*s23*c4+s1*s4)*s5+c1*c23*c5)*d_6)\dot\theta_1 -$$
$$((a_2c_2 + d_4s_{23} + a_3c_{23}) + d_6(c_{23}c_4s_5 + s_{23}c_5))s_1\dot\theta_2 - ((d_4s_{23} + a_3c_{23})s_1 +$$
$$+ d_6s_1(c_{23}c_4s_5 + s_{23}c_5))\dot\theta_3 + d_6(s_{23}s_1s_4s_5 - c_1c_4s_5)\dot\theta_4 - d_6(c_{23}s_1s_5 + c_1s_4c_5 +$$
$$+ s_1c_4c_5s_{23})\dot\theta_5$$

$${}^0v_6(z) = ((c_{23}c_5 - s_{23}c_4s_5)d_6 + d_4c_{23} - a_3s_{23} - a_2s_2)\dot\theta_2 + ((c_{23}c_5 - s_{23}c_4s_5)d_6 + d_4c_{23} -$$
$$a_3s_{23})\dot\theta_3 - s_5s_4c_{23}d_6 \quad + (c_{23}c_5c_4 - s_5s_{23})d_6\dot\theta_5$$

$${}^0w_6 =$$

$$\begin{bmatrix} s_1(\dot\theta_2 + \dot\theta_3) + c_1c_{23}\dot\theta_4 + (c_1s_{23}s_4 + s_1c_4)\dot\theta_5 + ((-c_1s_{23}c_4 + s_1s_4)s_5 + c_1c_{23}c_5)\dot\theta_6 \\ -c_1(\dot\theta_2 + \dot\theta_3) + s_1c_{23}\dot\theta_4 + (s_1s_{23}s_4 - c_1c_4)\dot\theta_5 - ((s_1s_{23}c_4 + c_1s_4)s_5 - s_1c_{23}c_5)\dot\theta_6 \\ \dot\theta_1 + s_{23}\dot\theta_4 - c_{23}s_4\dot\theta_5 + (c_{23}c_4s_5 + s_{23}c_5)\dot\theta_6 \end{bmatrix}$$

$${}^0_4J =$$

$$\begin{bmatrix} (a_2s_2 - a_1 + a_3s_{23} - d_4c_{23})s_1 & -(a_2c_2 + d_4s_{23} - a_3c_{23})c_1 & -(d_4s_{23} + a_3c_{23})c_1 & 0 \\ (a_1 - a_2s_2 + d_4c_{23} - a_3s_{23})c_1 & -(a_2c_2 + d_4s_{23} + a_3c_{23})s_1 & -(d_4s_{23} + a_3c_{23})s_1 & 0 \\ 0 & d_4c_{23} - a_3s_{23} - a_2s_2 & d_4c_{23} - a_3s_{23} & 0 \\ 0 & s_1 & s_1 & c_1c_{23} \\ 0 & -c_1 & -c_1 & s_1c_{23} \\ 1 & 0 & 0 & s_{23} \end{bmatrix}$$

$${}^0_3J = \begin{bmatrix} a_2s_1s_2 - a_1s_1 & -a_2c_1c_2 & 0 \\ a_1c_1 - a_2c_1s_2 & -a_2c_2s_1 & 0 \\ 0 & -a_2s_2 & 0 \\ 0 & s_1 & s_1 \\ 0 & -c_1 & -c_1 \\ 1 & 0 & 0 \end{bmatrix} \qquad {}^0_6J = \begin{bmatrix} J_{11} & J_{12} & J_{13} & J_{14} & J_{15} & J_{16} \\ J_{21} & J_{22} & J_{23} & J_{24} & J_{25} & J_{26} \\ J_{31} & J_{32} & J_{33} & J_{34} & J_{35} & J_{36} \\ J_{41} & J_{42} & J_{43} & J_{44} & J_{45} & J_{46} \\ J_{51} & J_{52} & J_{53} & J_{54} & J_{55} & J_{56} \\ J_{61} & J_{62} & J_{63} & J_{64} & J_{65} & J_{66} \end{bmatrix}$$

$J_{11} = $ (a2*s2 - a1 + a3*s23 - d4*c23)*s1 + d6*((s1*c4*s23 + c1*s4)*s5 - s1*c23*c5);

$J_{12} = $ ((-a2*c2 - d4*s23 - a3*c23) - d6*(c23*c4*s5 + s23*c5))*c1;

$J_{13} = $ c1*((-d4*s23 - a3*c23) - d6*(c23*c4*s5 + s23*c5));

$J_{14} = $ d6*(s1*c4*s5 + c1*s4*s5*s23);

$J_{15} = $ d6*(s1*c5*s4 - c1*c23*s5 - c1*c4*c5*s23);

$J_{16} = 0$;

$J_{21} = $ (a1 - a2*s2 + d4*c23 - a3*s23)*c1 +
+((-c1*s23*c4+s1*s4)*s5+c1*c23*c5)*d6;

$J_{22} = $ - ((a2*c2 + d4*s23 + a3*c23) + d6*(c23*c4*s5 + s23*c5))*s1;

J_{23} = - (d4*s23 + a3*c23)*s1 - d6*s1*(c23*c4*s5 + s23*c5);
J_{24} = d6*(s23*s1*s4*s5 - c1*c4*s5);
J_{25} = - d6*(c23*s1*s5 + c1*s4*c5 + s1*c4*c5*s23);
J_{26} = 0;

J_{31} = 0;
J_{32} = (c23*c5 - s23*c4*s5)*d6 + d4*c23 -a3*s23 -a2*s2;
J_{33} = (c23*c5 - s23*c4*s5)*d6 +d4*c23 -a3*s23;
J_{34} = - s5*s4*c23*d6;
J_{35} = (c23*c5*c4 - s5*s23)*d6;
J_{36} = 0;

J_{41} = 0;
J_{42} = s1;
J_{43} = s1;
J_{44} = c1*c23;
J_{45} = c1*s23*s4 + s1*c4;
J_{46}= (- c1*s23*c4 + s1*s4)*s5 + c1*c23*c5;

J_{51} = 0;
J_{52} = - c1;
J_{53} = - c1;
J_{54} = s1*c23;
J_{55} = s1*s23*s4 - c1*c4;
J_{56} = - ((s1*s23*c4 + c1*s4)*s5 - s1*c23*c5);

J_{61} = 1;
J_{62} = 0;
J_{63} = 0;
J_{64} = s23;
J_{65} = -c23*s4;
J_{66} = c23*c4*s5 + s23*c5;

Note: These calculations were made in *MatLab* using the symbolic Toolbox.

Figure 2.12 Linear and angular velocities, jacobian matrices ^0_3J, ^0_4J and ^0_6J

2.4 Singularities

If the objective is to use the differential kinematics equation (2.28) for simplicity and efficiency, then it's necessary to deal with the singularities of the jacobian. The differential kinematics equation maps the vector of joint velocities $\dot{q} = [\dot{q}_1 \quad \dot{q}_2 \quad \dot{q}_3 \quad \dot{q}_4 \quad \dot{q}_5 \quad \dot{q}_6]^T$ with the *end-effector* twist vector $V = [v^T \quad w^T]^T$. This mapping is seriously affected when the jacobian is rank-deficient (kinematics singularities), because in those situations the mobility of the robot is reduced, the inverse kinematics may show infinite solutions, and (because the jacobian determinant may take very small values near singularities) small task space velocities may cause very large joint velocities [2]. So, to control the robot manipulator it is necessary to find all singular configurations and design a scheme to identify a singular configuration approach.

In order to find all the singular points of the ABB IRB 1400 anthropomorphic industrial robot, which has a very simple kinematic structure, a scheme will be used that separates the arm singularities and the wrist singularities. By dividing the jacobian into four 3×3 blocks it can then be expressed as

$$
{}_6^0 J = \begin{bmatrix} J_{11} & J_{12} \\ J_{21} & J_{22} \end{bmatrix}
\tag{2.65}
$$

Now, looking to all the elements of J_{12} (Figure 2.12) it is clear that $\det(J_{12})$ vanishes making $d_6=0$. That is equivalent to choosing the origin of the *end-effector* frame coincident with the origin of axis 4 and 5, i.e., making $\mathbf{p_w} = \mathbf{p}$. Since singularities are a characteristic of the robot structure and do not depend on the frames chosen to describe kinematically the robot, this procedure is allowed. It's possible then to write

$$
\det(J) = \det(J_{11})*\det(J_{22})
\tag{2.66}
$$

The robot's singular configurations are the ones that make $\det(J) = 0$ which means from (2.66)

$$
\det(J_{11}) = 0 \quad \text{or} \quad \det(J_{22}) = 0
\tag{2.67}
$$

Solving the first equation leads to the so called *arm singularities* and solving the second leads to the *wrist singularities*.

Wrist Singularities
The wrist singularities can be found just by analyzing the structure of $\det(J_{22})$:

$$
\det(J_{22}) = \det\begin{pmatrix} z_4 & z_5 & z_6 \end{pmatrix} =
$$
$$
\det \begin{pmatrix} c_1c_{23} & c_1s_{23}s_4 - c_1c_4 & (s_1s_4 - c_1s_{23}c_5)s_5 + c_1c_{23}c_5 \\ s_1c_{23} & s_1s_{23}s_4 - c_1c_4 & -(s_1s_{23}c_4 + c_1s_4)s_5 + s_1c_{23}c_5 \\ s_{23} & -c_{23}s_4 & c_{23}c_4s_5 + s_{23}c_5 \end{pmatrix}
\tag{2.68}
$$

The above determinant is non-null if the column vectors of J_{22} (which correspond to z_4, z_5, and z_6) are linearly independent, i.e., the singular configurations are the ones that make at least two of them linearly dependent. Now, vectors z_4 and z_5 are linearly independent in all configurations, and the same occurs between z_5 and z_6. This conclusion is easy to understand looking to (2.68) and/or remembering that z_4

is perpendicular to z_5, and z_5 is perpendicular to z_6 in all possible robot configurations. A singular configuration appears when z_4 and z_6 are linearly dependent, i.e., when those axis align with each other, which means $s_5=0$ from (2.68). Consequently the wrist singular configurations occur when,

$$\theta_5 = 0 \quad \text{or} \quad \theta_5 = \pi \tag{2.69}$$

The second condition ($\theta_5 = \pi$) is out of joint 5 work range, and because of that is of no interest, i.e., the wrist singularities will occur whenever $\theta_5 = 0$.

Arm Singularities
The arm singularities occur when $\det(J_{11}) = 0$ making again $\mathbf{p} = \mathbf{p_w} \Rightarrow d_6 = 0$, i.e., when

$$\det \begin{pmatrix} (a_2s_2 - a_1 + a_3s_{23} - d_4c_{23})s_1 & -(a_2c_2 + d_4s_{23} - a_3c_{23})c_1 & -(d_4s_{23} + a_3c_{23})c_1 \\ (a_1 - a_2s_2 + d_4c_{23} - a_3s_{23})c_1 & -(a_2c_2 + d_4s_{23} + a_3c_{23})c_1 & -(d_4s_{23} + a_3c_{23})c_1 \\ 0 & d_4c_{23} - a_3s_{23} - a_2s_2 & d_4c_{23} - a_3s_{23} \end{pmatrix} = 0 \tag{2.70}$$

Solving (2.70) gives

$$-a_2(d_4c_3 - a_3s_3)(a_3s_{23} - d_4c_{23} + a_2s_2 - a_1) = 0 \tag{2.71}$$

which leads to the following conditions:

$$-a_3s_3 + d_4c_3 = 0$$

and/or

$$a_3s_{23} - d_4c_{23} + a_2s_2 - a_1 = 0 \tag{2.72}$$

The first condition leads to $\theta_3 = \text{arctg}\left(\dfrac{d_4}{a_3}\right)$. The elbow is completely stretched out and the robot manipulator is in the so called *elbow singularity*. This value of θ_3 is out of joint 3's work range, so it corresponds to a non-reachable configuration, and because of that is of no interest.

The second condition corresponds to configurations in which the origin of the wrist (origin of axis 4) lies in the axis of joint 1, i.e., lies in z_1 (note that z_1 is coincident with z_0). In those configurations, the position of the wrist cannot be changed by rotation of the remaining free joint θ_1 (remember that an anthropomorphic manipulator with a spherical wrist uses the anthropomorphic arm to position the spherical wrist, which is then used to set the orientation of the *end-effector*). The manipulator is in the so called *shoulder singularity*.

In conclusion, the arm singularities of the ABB IRB 1400 industrial robot are confined to all the configurations that correspond to a shoulder singularity, i.e., to configurations where $a_3s_{23} - d_4c_{23} + a_2s_2 - a_1 = 0$.

2.4.1 Brief Overview: Singularity Approach

As already mentioned, the solutions of the inverse kinematics problem can be computed from

$$\dot{q} = J^{-1}(\theta)V \tag{2.73}$$

solving (2.28) in order to $\dot{q}$. With this approach it's possible to compute the joint trajectories (q, $\dot{q}$), initially defined in terms of the *end-effector* wrist vector V and of the initial position/orientation. In fact, if q(0) is known it's possible to calculate:

$\dot{q}$ (t) from: $\dot{q}(t) = J^{-1}(\theta)V(t)$

and

q(t) from: $q(t) = q(0) + \int_0^t \dot{q}(\alpha)d\alpha$ (2.74)

Nevertheless, this is only possible if the jacobian is full rank, i.e., if the robot manipulator is out of singular configurations where the jacobian contains linearly dependent column vectors. In the neighborhood of a singularity, the jacobian inverse may take very high values, due to small values of det(J), i.e., in the neighborhood of a singular point small values of the velocity in the task space (V) can lead to very high values of the velocity in the joint space ($\dot{q}$).

The *singular value decomposition* (SVD) of the jacobian [3,8-10] is maybe the most general way to analyze what happens in the neighborhood of a singular point; also it is the only general reliable method to numerically determine the rank of the jacobian and the closeness to a singular point. With the inside given by the SVD of the jacobian, a *Damped Least-Square* scheme [9] can be optimized to be used in near-singular configurations. The *Damped Least-Square* (DLS) scheme trades-off accuracy of the inverse kinematics solutions with feasibility of those solutions: this trade-off is regulated by the damping factor ξ. To see how this works, let's define the DLS inverse jacobian by rewriting (2.28) in the form

$$(JJ^T + \xi^2 I)\dot{q} = J^T V \tag{2.75}$$

where ξ is the so-called damping factor. Solving (2.75) in order to $\dot{q}$ gives

$$\dot{q} = (JJ^T + \xi^2 I)^{-1} J^T V = J_{dls}^{-1} V \tag{2.76}$$

with J_{dls} being the *damped least-square* jacobian inverse. The solutions of (2.76) are the ones that minimize the following cost function [2,9,11]:

$$g(\dot{q}) = \frac{1}{2}(V - J\dot{q})^T (V - J\dot{q}) + \frac{1}{2}\xi^2 \dot{q}^T \dot{q} \tag{2.77}$$

resulting from the condition

$$\min_{\dot{q}} \left(\left\| V - J\dot{q} \right\|^2 + \xi^2 \left\| \dot{q} \right\|^2 \right) \tag{2.78}$$

The solutions are a trade-off between the *least-square* condition and the minimum norm condition. It is very important to select carefully the damping factor ξ : small values of ξ lead to accurate solutions but with low robustness to the singular or near-singular occurrences (= high degree of failure in singular or near-singular configurations), i.e., low robustness to the main reason to use the scheme. High values of ξ lead to feasible but awkward solutions.

To understand how to select the damping factor ξ, in the following the jacobian will be decomposed using the SVD technique. The SVD of the jacobian can be expressed as

$$J = U\Sigma V^T = \sum_{1}^{6} \sigma_i u_i v_i^T \tag{2.79}$$

where $\sigma_1 > \sigma_2 > ... > \sigma_r > 0$ (r = rank(J)) are the jacobian *singular values* (positive square roots of the eigenvalues of $J^T J$), v_i (columns of the orthogonal matrix V) are the so-called *right or input singular vectors* (orthonormal eigenvectors of $J^T J$) and u_i (columns of the orthogonal matrix U) are the so-called *left or output singular vectors* (orthonormal eigenvectors of JJ^T). The following properties hold:

$$R(J) = \text{span } \{u_1, ..., u_r\}^6$$
$$N(J) = \text{span } \{v_{r+1}, ..., v_6\}$$

The range of the jacobian R(J) is the set of all possible task velocities, those that could result from all possible joint velocities: $R(J) = \{V \in \Re^6 : V = J\dot{q}$ for all possible $\dot{q} \in \Re^6\}$. The first r input singular vectors constitute a base of R(J). So, if in a singularity the rank of the jacobian is reduced then one other effect of a singularity will be the decrease of dim[R(J)] by eliminating a linear combination of

---

[6] The span of $\{a_1, ... a_n\}$ is the set of the linear combinations of $a_1, ... a_n$.

task velocities from the space of feasible velocities, i.e., the reduction of the set of all possible task velocities.

The null space of the jacobian $N(J)$ is the set of all the joint velocities that produce a null task velocity at the current configuration: $N(J) = \{\dot{q} \in \Re^6 : J\dot{q} = 0\}$. The last (6-r) output singular vectors constitute a base of $N(J)$. So, in a singular configuration the dimension of $N(J)$ is increased by adding a linear combination of joint velocities that produce a null task velocity.

Using the SVD of the jacobian (2.78) in the DLS form of the inverse kinematics (2.75) results in

$$\dot{q} = \sum_1^6 \frac{\sigma_i}{\sigma_i^2 + \xi^2} v_i u_i^T V \tag{2.80}$$

The following properties hold:

$$R(J_{lds}) = R(J^\dagger)^7 = N^\perp(J)^8 = \text{span} \{u_1, \ldots, u_r\}$$
$$N(J_{lds}) = R(J^\dagger) = R^\perp(J)^9 = \text{span} \{v_{r+1}, \ldots, v_6\} \tag{2.81}$$

which means that the properties of the *damped least-squares* inverse solution are analogous to those of the pseudoinverse solution (remember that the inverse pseudoinverse solution gives a *least-square* solution with a minimum norm to equation (2.28)).

The damping factor has little influence on the components for which $\sigma_i \gg \xi$ because in those situations

$$\frac{\sigma_i}{\sigma_i^2 + \xi^2} \approx \frac{1}{\sigma_i} \tag{2.82}$$

i.e., the solutions are similar to the pure *least-square* solutions.

Nevertheless, when a singularity is approached, the smallest singular value (the r-th singular value) tend's to zero, the associated component of the solution is driven to zero by the factor $\frac{\sigma_i}{\xi^2}$ and the joint velocity associated with the near-degenerate components of the commanded velocity V are progressively reduced, i.e., at a singular configuration, the joint velocity along v_r is removed (no longer remains in the null-space of the jacobian as in the pure *Least-Square* solution) and the task

---

[7] $J^\dagger$ is the pseudoinverse jacobian.
[8] Orthogonal complement of the null space joint velocities.
[9] Orthogonal complement of the feasible space task velocities.

velocity along u_r becomes feasible. That is how the damping factor works; as a measure or indication of the degree of approximation between the damped and pure *least-square* solutions. Then a strategy [8], initially presented by [12], can be used to adjust ξ as a function of the closeness to the singularity. Based on the estimation of the smallest singular value of the jacobian, we can define a singular region and use the exact solution (ξ=0) outside the region and a damped solution inside the region. In this case, a varying ξ should be used (increasing as we approach the singular point) to achieve better performance (as mentioned the damped solutions are different from the exact solutions). The damping factor ξ can then be defined using the following law modified from [9]

$$\xi^2 = \begin{cases} 0 & \hat{\sigma}_6 \geq \varepsilon \\ \left(1 - \left(\dfrac{\hat{\sigma}_6}{\varepsilon}\right)^{2\eta}\right)\xi_{max}^2 & \hat{\sigma}_6 < \varepsilon \end{cases} \tag{2.83}$$

where ξ_{max}^2 and η are defined by the user to shape the solution to his needs, ε defines the size of the region and $\hat{\sigma}_6$ is the estimate of the smallest singular value. The estimate is done using a recursive algorithm originally presented at [13] and later extended by [14] to estimate not only the smallest singular value but also the second smallest singular value. This procedure avoids estimation inaccuracy due to the cross of the two smallest singular values, when the manipulator approaches both the wrist and the shoulder singularity. The algorithm is as follows:

Suppose we have estimates of the two last input singular vectors $\hat{v}_5$ and $\hat{v}_6$ with

$$\hat{v}_5 \approx v_5 \text{ and } \|\hat{v}_5\| = 1$$
$$\hat{v}_6 \approx v_6 \text{ and } \|\hat{v}_6\| = 1 \tag{2.84}$$

The estimate $\hat{v}_6$ is then use to compute $\hat{v}_6'$ from

$$\left(J^T J + \xi^2 I\right)\hat{v}_6' = \hat{v}_6 \tag{2.85}$$

Then the estimate $\hat{\sigma}_6^2$ is computed from

$$\hat{\sigma}_6^2 = \frac{1}{\|\hat{v}_6'\|} - \xi^2 \tag{2.86}$$

and the initial estimate $\hat{v}_6$ is updated using

$$\hat{v}_6 = \frac{\hat{v}_6'}{\left\|\hat{v}_6'\right\|} \tag{2.87}$$

The second smallest singular value is computed using the estimate $\hat{v}_6$ from,

$$\left[J^T J + \xi^2 I - \left(\hat{\sigma}_6^2 + \xi^2\right)\hat{v}_6\hat{v}_6^T\right]\hat{v}_5' = \hat{v}_5 \tag{2.88}$$

Then the estimate $\hat{\sigma}_5^2$ is computed from

$$\hat{\sigma}_5^2 = \frac{1}{\left\|\hat{v}_5'\right\|} - \xi^2 \tag{2.89}$$

and finally the initial estimate $\hat{v}_5$ is updated using

$$\hat{v}_5 = \frac{\hat{v}_5'}{\left\|\hat{v}_5'\right\|} \tag{2.90}$$

Special care should be taken with the numerical implementation of the DLS inverse kinematics solutions, to correct the numerical drift. Basically a feedback term can be used [2,9,15] by making

$$V = V_d + K.e = V_d + K \begin{pmatrix} p_d - p \\ \dfrac{1}{2}(n \times n_d + s \times s_d + a \times a_d) \end{pmatrix} \tag{2.91}$$

where K is a positive definite diagonal 6×6 matrix, p_d and p are the desired and actual position, and the orientation is defined in terms of the desired and actual (n, s, a) vectors of the *end-effector* frame.

Due to the increase of *end-effector* errors [11] in the neighborhood of a singularity by means of the near-degenerate components of *end-effector* velocity, the matrix K should be corrected using $K=\rho K_0$, where K_0 is a diagonal constant matrix and ρ is the correcting factor. Now, inside a singular region we should use K=0 because in some situations the resulting joint velocities can drive the manipulator to reach the joint limits, even if eventually the error will approach zero. When the manipulator is sufficiently away from a singularity, we should have $\rho=1$. So, generally we define ρ as

$$\rho = \begin{cases} 0 & \sigma_6 \le \varepsilon \\ \dfrac{(\sigma_6 - \varepsilon)^2}{(n-1)^2 \varepsilon^2} & \varepsilon < \sigma_6 < n\varepsilon \\ 1 & \text{otherwise} \end{cases} \qquad (2.92)$$

where **n** is defined by the user based on self-experience and on test results with a particular robotic manipulator setup.

2.5 Position Sensing

The IRB1400 uses resolvers [16-19] as position sensors. The drive unit used at this robot (manufactured by *ELMO AB* for *ABB Robotics*), includes a PM AC synchronous motor, both current feedback devices, a brake, and a brushless resolver, all assembled at factory ,i.e., they come in one piece [20].

A brushless resolver consists of a stator, a rotor and a rotary transformer. The stator and rotor windings are distributed in a way that the magnetic flux is distributed as a sine wave of the angle of rotation (perfect resolver). The output of a resolver is therefore an AC voltage in accordance with the angular position of the shaft. This type of position sensor is characterized by its high accuracy output, maintenance free brushless design, and immunity to noise, vibration, and shock. Other characteristics introduced by highly automated manufacturer production facilities include homogeneity in accuracy, transformation ratio, phase-shift, etc.

These characteristics significantly reduce major sources of error such as:

1. Amplitude imbalance due to different amplitudes of the resolver output signals
2. Imperfect quadrature due to phase-shift
3. Inductance harmonic error due to imperfect inductance profiles, i.e., the inductance profiles do not follow perfect sine wave as consequence of imperfect sinusoidal winding

Two types of resolvers can be considered (Figure 2.13): *Brushless Amplitude Output Resolvers* (BAOR) and *Brushless Phase-Shift Output Resolvers* (BPOR)[10]. Resolvers of type BAOR are excited by an AC voltage to the rotor winding and the output is obtained from the stator windings in the form sine and cosine voltages proportional to the rotation angle θ. Resolvers of the type BPOR are excited by sine and cosine voltages to the stator windings and the output is obtained from the

---

[10] *Tamagawa Seiki Co. LTD.* names these resolvers as BRX and BRT, respectively.

rotor winding in the form of a sine voltage with phase-shifted in proportion to the rotation angle θ.

The IRB 1400 uses BAOR type resolvers from the Japanese manufacturer Tamagawa Seiki Co. LTD. [19,20].

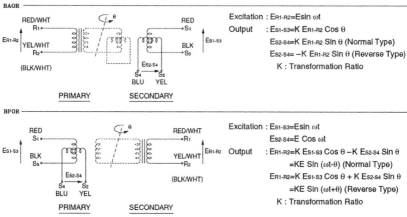

Figure 2.13 Types of resolvers

The use of a resolver implies the availability of a *resolver to digital converter* (RDC) and processing circuit [21-23]. The RDC is used to track and convert resolver signals to a digital parallel binary word, generally using a ratiometric tracking conversion method that improves noise immunity and tolerance to lead length (important when the converter is remote from the resolver). The RDC circuit uses an RDC along with the necessary interface and signal conditioning circuitry. Because noise can degrade significantly the accuracy of the measurement, special care must be taken with the driving lines from and to the resolvers: the use of shielded twisted pair cabling and isolation amplifiers may be needed.

The basic functional diagram of an RDC is presented in Figure 2.14, where it is used data relative to the *Analog Devices* RDC model AD2S80A. The converter works as a type II closed-loop system with the angle ϕ as a control variable (this angle is the current converter estimate of the angle θ).

Generally, the converter's functioning can be described as follows: First the inputs (resolver outputs E_{S2-S4} and E_{S1-S3}) are multiplied by $\cos(\phi)$ and $\sin(\phi)$, respectively, at the *ratio multiplier*. Then the difference between the signals is computed giving the ratio multiplier output AC error signal $E_{ac} = A_1.K.E.\sin(\theta - \phi).\sin(\omega t)$, where A_1 is the ratio multiplier gain (fixed at 14.5 for AD2S80A). Second, this error signal is synchronously demodulated at the *phase sensitive demodulator* (PSD), using the resolver excitation frequency as a demodulation reference, leaving the error signal

$E_{PSD} = A_1.K.E\sin(\theta - \phi)$. The output of the demodulator is a DC voltage proportional to the RMS value of the demodulator input: $\dfrac{\pm\sqrt{2}}{\pi} * \text{Demodulator\_Input}_{RMS}$ (for sinusoidal input signals in phase or antiphase with the reference signal). Before entering the PSD, the signal passes over an *HF filter* (with components selected by the user) to remove any DC offset voltage. Then the PSD output passes through the *integrator* (with components selected by the user), whose output signal (proportional to the velocity of the resolver) is fed to the *voltage controlled oscillator* (VCO). The VCO integrates the velocity signal and compares the resulting signal with the minimum DC voltage resolution (uses two comparators for positive and negative voltages, meaning rotation in the positive direction or in the negative direction) and updates the up/down counter by producing the counter clock and direction signal. The value of the internal latch used to interface with the user is also updated with the counter value. An RDC works similarly to a *successive approximation* type *analog to digital converter*.

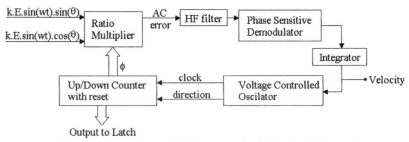

Figure 2.14 Resolver to digital converter basic functional diagram

The RDC returned digital value is generally a 12, 14, or 16-bit binary number containing the actual rotation angle. This angle should be mapped to the robot's join space. For that, the following guidelines should be used:

1. Choose an angle data format, i.e., degrees, radians
2. Account for the resolver offset[11], i.e., the resolver reading when the manipulator is in the home position. At that point, we should have *number_of_rotations* = 0 and *actual_angle* = 0

A complete RDC circuit implementation should also save the total number of rotations in an 8-bit *up-down* counter/register. In essence, the circuit should give the rotation angle of the motor in the actual rotation and the total number of rotations already performed.

---

[11] Usually these values are measured by the robot manufacturer and printed on the robot or in the robot documentation.

2.6 Actuators: Motors

Generally the actuators used to move the joints of any industrial robot are motors, usually DC permanent magnet (PM) motors or AC PM motors. Other motors can be used, including pneumatic or hydraulic servo motors. The IRB 1400 uses three-phase synchronous AC PM motors, with six poles (axes 1-3) and four poles (axes 4-6), manufactured by *Elmo AB* – Sweden.

The three-phase synchronous AC PM motor rotating magnetic field is obtained by making a three-phase current to flow in the stator coil (Figure 2.15), which has a sinusoidal distribution. So, a brushless sine wave PM AC synchronous motor is obviously not mechanically commutated (there are no brushes) but instead the commutation is done by acting on the three-phase current signals. Nevertheless, the commutation position of the motor should be retained, i.e., the resolver reading when the motor is at the electrical home position (electrical $0°$ position) - this value is called the commutation offset (COMMOFF).

The usual procedure to find the commutation offsets is as follows:
1. Turn the motor to the commutating position by feeding a positive constant current to the motor
2. Feed the resolver with the necessary excitation signal (4kHz and 5 V_{rms} for IRB 1400 drives)
3. Adjust the resolver to $+90°$ ($\pm 0,5°$), i.e, turn to the maximum value on coil Y of the resolver with the same phase as the 5V feeding signal. At that point we should have:
    Voltage across coil X = 0V
    and
    Voltage across coil Y = input voltage * transformation ratio

The value of the rotation angle (90 degrees) is the commutation offset. This procedure is used with the IRB 1400 drives, so that is why the COMMOFFS are constant for all drives (1.570800 radians). For some older robots, like the ABB IRB 2000 (up to model M90), the motor and the resolver are separate parts, assembled together by the manufacturer without following the above referred procedure. So, the COMMOFFS are different for all drives. The values are obtained at factory and printed on the robot or in the documentation; nevertheless, these values can be updated using the robot controller.

A full description of a three-phase synchronous sine wave PM motor can be found in:

1. *Design of Brushless Permanent-Magnet Motors*, Herdershot Jr., Magna Physics Publishing and Clarendon Press, Oxford, 1995, Chapters 6 and 7
2. *Electric Drives and their Controls*, R.M. Crowder, Clarendon Press, Oxford, 1995, Chapter 5, section 5.3

Nevertheless, a brief overview is presented here.

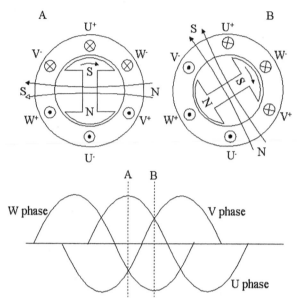

Figure 2.15 Three-phase synchronous motors and current signals

Considering β as the angle between rotor magnet north axis and the stator windings axis, it can be shown [17] that the motor torque is

$$T \propto \sin(\beta) \tag{2.93}$$

Consequently, the angle β must be kept at 90° in order to maximize the torque, which is done by phasing the current waveforms relative to the actual rotor position. To ensure that the ampere-conductor distribution remains in synchronism with the rotor's magnetic field, the stator supply frequency (f) must be equal to the rotor angular velocity (w_s), $w_s = 2.\pi.f$, which is related to the mechanical angular velocity of the motor (w_m) by $w_m = w_s/p$, where p is the number of the motor pole pairs. In order to keep the torque angle constant, i.e., to keep the ampere-distribution north axis in synchronism with the rotor north axis (displaced by 90°), a high-performance and precise sensor should be used (generally a resolver).

With this type of control action the motor follows the equation

$$\text{Torque} = \text{Flux} * \text{Current} \tag{2.94}$$

For this type of motors, the flux is constant, sinusoidally distributed in space, and the generated EMF varies sinusoidally in each phase. The overall torque-speed characteristic is presented in Figure 2.16. The maximum torque can be maintained up to the base speed. After that, it is still possible to increase the velocity by changing β but the motor enters the field-weakening mode and any increase in speed is done at the expense of the peak torque.

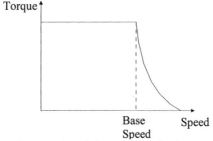

Figure 2.16 Torque-Speed characteristic of a sine wave motor

The "natural" relations for the back-EMF (E) and for the torque (T), used for a DC square wave motor still hold for a sine wave motor, i.e.,

$$T = k_t * I$$
$$E = k_e * w_m \tag{2.95}$$

but now with $\dfrac{k_t}{k_e} = \dfrac{\sqrt{3}}{2} \neq 1$.

The torque constant (k_t) and the back-EMF constant (k_e) can be measured using the following equations:

$$k_e = \frac{\hat{e}_{LL}}{w_m} \text{ (V-s/rad)} \tag{2.96}$$

where $\hat{e}_{LL}$ is the peak line-line voltage and w_m is the mechanical angular velocity.

$$k_t = \frac{T}{\hat{i}} \text{ (Nm)} \tag{2.97}$$

where $\hat{i}$ is the peak line current when the motor is in normal operation, measured using a current sensor connected to measure the phase current directly and then displayed in an oscilloscope.

It is also possible to write

$$T.w_m = k_t * \hat{i} * \frac{\hat{e}_{LL}}{k_e} = \frac{\sqrt{3}}{2} * \hat{e}_{LL} * \hat{i} = \frac{\sqrt{3}}{2} * \sqrt{2} * E_{RMS} * \sqrt{2} * I_{RMS}$$

(2.98)

$$= \sqrt{3} * E_{RMS} * I_{RMS} = \text{Electrical} - \text{Mechanical Power Conversion}$$

and,

$$T = \frac{\sqrt{3} * E_{RMS} * I_{RMS}}{w_m} = k_t * I_{RMS} \Rightarrow k_t = \frac{\sqrt{3} * E_{RMS}}{2\pi * \dfrac{Vol_{RPM}}{60}}$$

(2.99)

2.6.1 Motor Drive System

In this section, the main circuits necessary to drive a three-phase AC synchronous PM motor are briefly presented. As already mentioned, a brushless AC PM motor requires alternating sine wave phase currents, because the motor is designed to generate sinusoidal back-EMF. The power electronic control circuit is very simple and uses some control strategy[12] to achieve torque, smooth speed, and accurate control, keeping the current to a safe value. In order to obtain sine wave phase currents, the power supply (DC voltage) must be switched on and off at high frequency, under the control of a current regulator that forces the power transistors to switch on and off in a way that the average current is a sine wave. Basically, the sine wave reference signals could just be applied directly to the power transistors, after appropriate power amplification. However, that means using the power transistors in the proportional or linear region, which will increase the operating temperature due to the high power loss. The power loss is reduced by switching the transistors on and off by comparing the sine wave reference with a high frequency triangular carrier wave (PWM - *pulse width modulation* circuit). The frequency and amplitude of the triangular wave are kept constant. The comparator switches on the transistors when the values of the reference sine wave exceed those of the triangular wave; and switches them off when the inverse situation occurs (Figure 2.17). The duty ratio is then increased and decreased by the sine wave, centered by 50%. This procedure leads to a average sine wave output, because the output of the inverter feeding the power transistors is 0V when the duty ratio is 50%.

Special care should be taken in selecting the carrier frequency, because the power loss increases with increasing frequency and the motor speed response decreases with decreasing frequency. Torque and current ripples appear more frequently at higher frequencies as well.

---

[12] A set of rules that determine when the power transistors are switched on and off

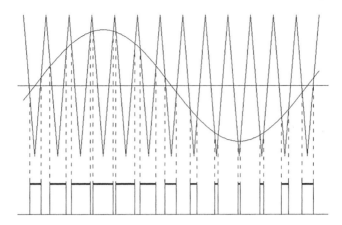

Figure 2.17 PWM basic functioning

The basic power electronic circuit to control a sine wave three-phase AC PM motor is the full-bridge circuit. The transistors used in the circuit must have very low turn-on and turn-off switching times (of the order of nanoseconds) and some other properties summarized as follows:

1. Zero on-state forward voltage drop, to minimize losses and maximize available "voltage" to force current into the motor
2. Zero leakage current in the off state, to minimize losses because a power transistor usually has high voltages across it when it is off, so even a small leakage current can produce high losses in the transistor's off state
3. High forward-blocking capability that should be higher than the supply voltage by a safety margin (usually 30%). The reverse-blocking capability is generally a margin of the forward-blocking, usually because the power transistors are reverse-protected by appropriately connected diodes
4. High dv/dt capability, because modern power transistors are MOS-gated, with capacitive input impedance at the gate, which make's them sensitive to spurious turn-on when the gate is subjected to a high dv/dt. High dv/dt immunity is then desirable, but nevertheless a safe procedure is to drive the gate from a low impedance source/sink
5. High di/dt capability, to prevent current-crowding effects and second breakdown the di/dt capability must be high
6. High-speed switching, from transistors to minimize switching losses and also from the power diodes, because the commutation of inductive current from a transistor branch to a diode branch is the most important way to protect against destructive transient voltages

The full bridge circuit is presented in Figure 2.18 for two popular phase windings: eye and delta [17]. Figure 2.19 shows line current waveforms for three-phase sine wave motors, including transistor states and current paths.

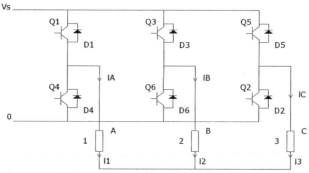

Eye connected

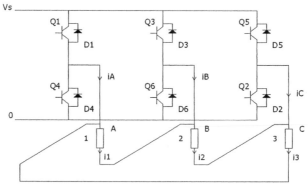

Delta connected

Figure 2.18 Full bridge circuit for eye and delta connected windings

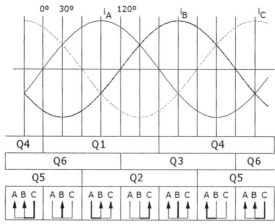

Figure 2.19 Line current waveforms for a sine wave motor, including transistor states and current paths

A general control system for a sine wave three-phase brushless motor is presented in Figure 2.20: includes a PWM circuit, over current (due to motor stall or short circuits) protection, a filter to damp DAC steps, a current controller (usually a PI controller designed to drive the motor current to the desired value) and a sine wave generator. Synchronization is achieved by changing current references in accordance with motor position.

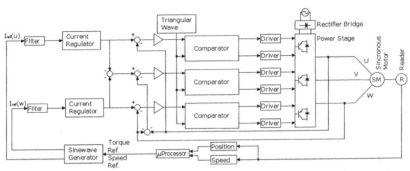

Figure 2.20 Block diagram of a general control system for a brushless synchronous three-phase sine wave motor

2.7 Dynamics

Dynamics deals with mapping forces exerted on the robot's parts as well as with the motion of the robot, i.e., its joint positions, velocities, and accelerations. This mapping is achieved using a set of mathematical equations, based on some specified dynamic formulation that describes the dynamic behavior of the robot manipulator, i.e., its motion. Those sets of equations constitute the dynamic model of the robot manipulator. The dynamic model can be used to simulate and control the robot manipulator, i.e., the dynamic model provides the means to compute the joint positions, velocities, and accelerations starting from the joint torques (*direct dynamics*), and the means to compute the joint torques using the joint positions, velocities, and accelerations (*inverse dynamics*).

The dynamic model is obtained starting from well known physical laws like the *Newtonian* mechanics and the *Lagrange* mechanics [6,24]. Several different dynamic formulations for robot manipulators were developed: *Lagrange-Euler*, *Newton-Euler*, *D'Alembert*, ... [1-3,7]. Nevertheless, they are equivalent to each other because they define the same physical phenomenon, i.e., the dynamics of rigid bodies assembled together to constitute a robot. Obviously, the structure of the motion equations is much different because each formulation was developed to achieve different objectives such as computation efficiency, simplicity to analyze and/or to simulate the structure, etc.

In this section, the dynamic model of the ABB IRB 1400 industrial robot will be briefly summarized using the Newton-Euler dynamic formulation. In the process, the other dynamic formulations are presented and briefly discussed.

2.7.1 Inertia Tensor and Mass Distribution

The mass distribution of a rigid body may be characterized by its inertial mass, for the case of one degree of freedom motions, and by its first moment of inertia, for simple rotations, i.e., rotations about a single axis. If there is more than one axis of rotation, the above properties are no longer suitable to characterize the mass distribution of the moving rigid body [6,24]. This is the case of a rigid robot manipulator, which is made by a series of rigid bodies, whose motion is 3-dimensional and therefore an infinite number of rotation axes is possible. The concept of *inertia tensor* is used in this case, which can be considered as a generalization of the concept of moment of inertia. If $\rho(x,y,z)$ is the mass density of a rigid body, then the inertia tensor may be defined as

$$I = \iiint \rho(r^2 \mathbf{1} - rr)dv \qquad (2.100)$$

where $\mathbf{1}$ is a unity tensor. The inertia tensor is a 3×3 matrix expressed in terms of some frame {A}

$$^A I = \begin{bmatrix} I_{xx} & I_{yx} & I_{zx} \\ I_{xy} & I_{yy} & I_{zy} \\ I_{xz} & I_{yz} & I_{zz} \end{bmatrix} \qquad (2.101)$$

where the diagonal elements are the moments of inertia about the axes x, y and z of frame {A}

$$I_{xx} = \iiint \rho(y^2 + z^2)dv$$
$$I_{yy} = \iiint \rho(z^2 + x^2)dv$$
$$I_{zz} = \iiint \rho(x^2 + y^2)dv \qquad (2.102)$$

and the other elements (non-diagonal) are the products of inertia

$$I_{xy} = I_{yx} = -\iint \rho xy dv$$
$$I_{yz} = I_{zy} = -\iint \rho yz dv$$
$$I_{zx} = I_{xz} = -\iint \rho zx dv \qquad (2.103)$$

2.7.1.1 Important Results [6]
Next some important results will be presented, considering that the frame associated to the rigid body is {B} and the inertial frame is {A}.

Suppose that I is the inertia tensor of the rigid body expressed in terms of some reference frame. The moment of inertia about any axis of rotation **n** (different from any of the rigid body symmetry axes) with the same origin of the reference frame is

$$I_n = n^T I . n \qquad (2.104)$$

Extension of the Parallel Axis Theorem This theorem is used here to compute the inertia tensor variation with linear motions of the reference frame. Suppose that {C} is the frame associated with the rigid body center of mass, {G} is some frame obtained from {C} by linear motion, and CP is the position vector of the center of mass expressed in terms of {G}. Then

$$I_G = I_C + M \left({}^C P^T {}^C P I_3 - {}^C P {}^C P^T \right) \qquad (2.105)$$

where ${}^C P = (x_c, y_c, z_c)^T$ and I_3 is a 3×3 identity matrix.

If the rigid body is rotating, the inertia tensor expressed in terms of {A} $^A I$ is also varying with time, but the inertia tensor expressed in terms o {B} $^B I$ remains constant (remember that {B} is the frame associated with the rigid body). If the inertia tensor $^B I$ is known then

$$^A I = {}_B^A H . {}^B I . {}_B^A H^T \qquad (2.106)$$

where ${}_B^A H$ is the transformation matrix from {B} to {A}.

The reference frame associated with each rigid body must be set to in a way that the products of inertia become null. The axes of that frame are named *primary axes* of the rigid body. The eigenvalues of the inertia tensor are the so-called rigid body *primary moments of inertia*. There are some systematic methods to compute the primary axis of inertia of any rigid body [6,24].

Any rigid body plane of symmetry is perpendicular to one primary axis.

Each symmetry axis of the rigid body is a primary axis. The plane of symmetry perpendicular to that axis is a *primary plane* associated with a degenerated primary moment of inertia.

2.7.2 Lagrange-Euler Formulation

Here we briefly introduce the *Lagrange-Euler* formulation. To use this formulation, it is required to develop equations for the robot manipulator's kinetic energy and potential energy. The kinetic energy of link (i) is given by

$$k_i = \frac{1}{2} m_i V_{C_i}^T V_{C_i} + \frac{1}{2} {}^i w_i^T . {}^{C_i} I_i {}^i w_i \tag{2.107}$$

where the first term results from the linear velocity of the center of mass of link (i), and the second term is due to the angular velocity of the same link. The robot manipulator's total kinetic energy is then given by

$$K = \sum_{i=1}^{6} k_i \tag{2.108}$$

The potential energy of link (i) may be written as

$$u_i = m_i . {}^0 g^T . {}^0 P_{C_i} + u_{ref_i} \tag{2.109}$$

where ${}^0 g$ is the gravity acceleration vector, ${}^0 P_{C_i}$ is the position vector of the center of mass of link (i) expressed in terms of frame {0} and u_{ref_i} is a constant that expresses the potential energy in terms of an arbitrary origin. The total potential energy of the robot manipulator is given by

$$U = \sum_{i=1}^{6} u_i \tag{2.110}$$

The *Lagrange* equation is then

$$L = K - U \tag{2.111}$$

where K and U are obtained form (2.100) and (2.110). It follows that the motion equations of the robot manipulator can then be obtained using the Lagrange equation

$$\tau = \frac{d}{dt} \frac{\partial L}{\partial \dot{\theta}} - \frac{\partial L}{\partial \theta} \tag{2.112}$$

where τ is the joint torque vector.

Recently [4], recursive equations based on the *Lagrange-Euler* equations have been developed. The resulting equations are computationally more efficient. Nevertheless, the recursive nature destroys the equation's structure which is a major drawback for the design and development of new control laws, and the *Newton-Euler* recursive equations remain the most efficient.

2.7.3 D'Alembert Formulation

This is basically a *Lagrange* dynamic formulation based on the *D'Alembert* principle. As mentioned before, the *Lagrange-Euler* formulation is simple but computationally inefficient, and the *Newton-Euler* formulation is compact with a recursive non-structured nature and is computationally very efficient. To obtain a recursive and computationally efficient set based on the Lagrange mechanics, a vector representation along with the use of rotation matrices is used to develop the kinetic and potential energy equations. The same procedure used in the *Lagrange-Euler* formulation is then used to compute the motion equations. This procedure is known as *D'Alembert* formulation, and is a generalization of the *Lagrange-Euler* and *Newton-Euler* formulations [7].

2.7.4 Newton-Euler Formulation

The *Newton-Euler* formulation will be used to obtain the dynamic equations of the ABB IRB 1400 industrial robot and in the process explained in some detail. We will also compare this to the other dynamic formulations.

If the joint positions, velocities, and accelerations of the robot manipulator are known, along with the kinematics and mass distribution, then we should be able to compute the required joint moments. On the other hand, if the joint torques is known, along with the inverse kinematics and the robot mass distribution, we should be able to compute the joint positions.

The *Newton-Euler* dynamic formulation is a set of recursive equations, divided in two groups: *forward recursive equations* and *inverse recursive equations*.

Forward Recursive Equations
This set of equations is used to compute ("*propagate*") link velocities and accelerations from link to link, starting from link 1 (the first link).

Angular Acceleration Computation
Using equations (2.50) and (2.51) gives

$$^{i+1}w_{i+1} = {}^{i+1}_iR.\,^i\dot{w}_i + {}^{i+1}_iR.\,^iw_i \times \dot{\theta}_{i+1}.\,^{i+1}Z_{i+1} + \ddot{\theta}_{i+1}.\,^{i+1}Z_{i+1} \qquad (2.113)$$

for the angular acceleration of link (i+1) expressed in terms of (i+1).

Linear Acceleration Computation
Using equations (2.52) and (2.53) gives

$$^{i+1}\dot{v}_{i+1} = {}^{i+1}_i R \left[{}^i\dot{w}_i \times {}^i P_{i+1} + {}^i w_i \times ({}^i w_i \times {}^i P_{i+1}) + {}^i \dot{v}_i \right] \qquad (2.114)$$

for the linear acceleration of link (i+1) expressed in terms of (i+1).

Linear Acceleration Computation at the Link Center of Mass
Using again equations (20) and (25) results,

$$^i \dot{v}_{C_i} = {}^i\dot{w}_i \times {}^i P_{C_i} + {}^i w_i \times ({}^i w_i \times {}^i P_{C_i}) + {}^i \dot{v}_i \qquad (2.115)$$

where $\{C_i\}$ is the reference frame associated with the center of mass of link (i), and having the same orientation of $\{i\}$.

Gravity effects
The gravity effects can be included in the above equations by making

$$^0 \dot{v}_0 = G \qquad (2.116)$$

where $G = \{g_x, g_y, g_z\}^T$ is the gravity acceleration vector with $|G| = 9{,}8062$ m/s^2. This is equivalent to consider that the robot manipulator has a linear acceleration of one G, pointing up, which produces the same effect on the robot links as the gravity acceleration.

Using the above equations (2.113)-(2.115), the *Newton* equation (2nd law) and the *Euler* equation, it's possible to compute the total force and moment at the center of mass of each link:

$$^{i+1}F_{i+1} = m_{i+1} \cdot {}^{i+1}\dot{v}_{C_{i+1}} \qquad (2.117)$$

$$^{i+1}N_{i+1} = {}^{C_{i+1}}I_{i+1} \cdot {}^{i+1}\dot{w}_{i+1} + {}^{i+1}w_{i+1} \times {}^{C_{i+1}}I_{i+1} \cdot {}^{i+1}w_{i+1} \qquad (2.118)$$

Note:
Newton Equation (2nd law) - The total force applied to a rigid body of mass m and centre of mass acceleration $\dot{v}_C$, is given by $F = m \cdot \dot{v}_C$.

Euler Equation - Consider a rigid body of mass m, angular velocity w, and angular acceleration $\dot{w}$. The total moment N starting the body in motion is given by $N = {}^C I \dot{w} + w \times {}^C I w$, where $^C I$ is the rigid body inertia tensor expressed in terms of reference frame associated with the body's center of mass.

Backward Recursive Equations
This set of equations is used to compute ("*propagate*") link forces and moments from link to link, starting at the last link.

Computation of Links Forces and Moments
Taking

$$f_i = \text{force applied at link (i) by link (i-1)};$$
$$n_i = \text{moment in link (i) due to link (i-1)};$$

the force balancing on link (i) can be expressed as

$$^iF_i = {}^if_i - {}^{i+1}_iR.^{i+1}f_{i+1} \tag{2.119}$$

and the moment balancing in the center of mass of link (i) can be expressed as

$$^iN_i = {}^in_i - {}^in_{i+1} + (-{}^iP_{C_i}) \times {}^if_i - ({}^iP_{i+1} - {}^iP_{C_i}) \times {}^if_{i+1} \tag{2.120}$$

Using (2.119) in (2.120) gives

$$^iN_i = {}^in_i - {}^{i+1}_iR.^{i+1}n_{i+1} + -{}^iP_{C_i} \times {}^iF_i - {}^iP_{i+1} \times {}^{i+1}_iR.^if_{i+1} \tag{2.121}$$

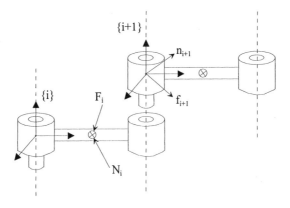

Figure 2.21 Forces and torques applied to the joints

Rewriting (2.119) and (2.121) in a way that their recursive nature becomes more evident results in

$$^if_i = {}^{i+1}_iR.^{i+1}f_{i+1} + {}^iF_i \tag{2.122}$$

$$^in_i = {}^iN_i + {}^{i+1}_iR.^{i+1}n_{i+1} + {}^iP_{C_i} \times {}^iF_i + {}^iP_{i+1} \times {}^{i+1}_iR.^{i+1}f_{i+1} \tag{2.123}$$

To obtain the joint moments we just need to project over the Z axis the already computed moment ${}^{i}n_i$, i.e.,

$$\tau_i = {}^{i}n_i^{T} \cdot {}^{i}Z_i \tag{2.124}$$

Contact Forces

The contact forces and moments (contact wrench) can be included in the model by putting,

$$\begin{pmatrix} {}^{N+1}f_{N+1} \\ {}^{N+1}n_{N+1} \end{pmatrix} = \text{Contact wrench} \neq 0 \tag{2.125}$$

where N is the number of degrees of freedom of the robot manipulator.

2.7.5 Dynamic Parameters

There is a number of parameters that are needed to compute the dynamic model (dynamic parameters). The minimum set of parameters is called the base dynamic parameters, and its identification can reduce significantly the computational load of the dynamic model (by 50%). If we take a closer look at the equations developed for the kinematics energy and for the potential energy of link (i), it is easy to verify that they are linear with respect to some dynamic parameters: the link mass, the six elements of the link inertia tensor, and the three components of the link's first moments of inertia. Some other dynamic parameters must also be included, namely the ones related with joint actuation. The joint torque is given by

$$\tau = \tau_m + \tau_v + \tau_f + \tau_g + \tau_\alpha + \tau_e \tag{2.126}$$

where $\tau_m = M(\theta)\ddot{\theta}$ is the torque due to the inertia of the robot manipulator, τ_v is the torque due to the centrifugal and coriolis forces, τ_f is the torque due to the friction forces, τ_g is the torque due to the gravity force, τ_α is the torque resulting from non-modeled forces and τ_e is the torque due to external contact forces.

Now, τ_m can be written as $\tau_m = \tau_{mr} + \tau_{mm}$, where τ_{mr} is the torque due to the robot manipulator inertia (not including the motor drive) and τ_{mm} is the torque due to the motor inertia itself. We may express τ_{mm} as

$$\tau_{mm} = I_m \cdot \ddot{\theta} = \begin{bmatrix} I_{m_1} & 0 & \cdots & 0 \\ 0 & I_{m_2} & \cdots & \\ \cdots & \cdots & \cdots & \cdots \\ 0 & 0 & \cdots & I_{m_n} \end{bmatrix} \cdot \begin{bmatrix} \ddot{\theta}_1 \\ \ddot{\theta}_2 \\ \cdots \\ \ddot{\theta}_n \end{bmatrix} \tag{2.127}$$

where I_m is the rotor's moment of inertia and n is the number of degrees of freedom.

The friction torque may be given by

$$\tau_f = F_s.\text{sgn}(\dot{\theta}) + F_v.\dot{\theta} = \begin{bmatrix} F_{s_1} & 0 & \cdots & 0 \\ 0 & F_{s_2} & \cdots & \\ \cdots & \cdots & \cdots & \cdots \\ 0 & 0 & \cdots & F_{s_n} \end{bmatrix}.\text{sgn}\begin{bmatrix} \dot{\theta}_1 \\ \dot{\theta}_2 \\ \cdots \\ \dot{\theta}_n \end{bmatrix} + \begin{bmatrix} F_{v_1} & 0 & \cdots & 0 \\ 0 & F_{v_2} & \cdots & \\ \cdots & \cdots & \cdots & \cdots \\ 0 & 0 & \cdots & F_{v_n} \end{bmatrix}.\begin{bmatrix} \dot{\theta}_1 \\ \dot{\theta}_2 \\ \cdots \\ \dot{\theta}_n \end{bmatrix} \qquad (2.128)$$

where the first term refers to the coulomb friction and the second to the viscous friction.

In conclusion, I_{m_i}, F_{s_i} and F_{v_i} are also dynamic parameters to take into account, i.e., the all number of dynamic parameters is thirteen:

$$\pi = \begin{pmatrix} I_{xx_i} & I_{yy_i} & I_{zz_i} & I_{xy_i} & I_{xz_i} & I_{yz_i} & m_i r_{ix} & m_i r_{iy} & m_i r_{iz} & m_i & I_{m_i} & F_{s_i} & F_{v_i} \end{pmatrix}$$
$$(2.129)$$

The basic Newton-Euler recursive algorithm resumed in the following form:

Forward recursive equations
Initial conditions
$^0w_0 = 0$; $^0\dot{w}_0 = 0$; $^0\dot{v}_0 = ^0\ddot{p}_0 = ^0\hat{g} = \begin{pmatrix} 0 & 0 & g \end{pmatrix}^T$, with g = - 9,8062 m/s².

For i = 1 to 5,
$$^{i+1}w_{i+1} = ^{i+1}_iR.^i\dot{w}_i + ^{i+1}_iR.^iw_i \times \dot{\theta}_{i+1}.^{i+1}Z_{i+1} + \ddot{\theta}_{i+1}.^{i+1}Z_{i+1}$$
$$^{i+1}\dot{v}_{i+1} = ^{i+1}_iR\left[^i\dot{w}_i \times ^iP_{i+1} + ^iw_i \times (^iw_i \times ^iP_{i+1}) + ^i\dot{v}_i\right]$$
$$^i\dot{v}_{C_i} = ^i\dot{w}_i \times ^iP_{C_i} + ^iw_i \times (^iw_i \times ^iP_{C_i}) + ^i\dot{v}_i$$
$$^{i+1}F_{i+1} = m_{i+1}.^{i+1}\dot{v}_{C_{i+1}}$$
$$^{i+1}N_{i+1} = ^{C_{i+1}}I_{i+1}.^{i+1}\dot{w}_{i+1} + ^{i+1}w_{i+1} \times ^{C_{i+1}}I_{i+1}.^{i+1}w_{i+1}$$

Backward recursive equations
Initial conditions
$$\textit{End-effector} \text{ wrench} = \begin{pmatrix} ^{N+1}f_{N+1} \\ ^{N+1}n_{N+1} \end{pmatrix}$$

For i = 6 to 1,
$$^if_i = ^i_{i+1}R.^{i+1}f_{i+1} + ^iF_i$$

$$^i n_i =^i N_i +_{i+1}^i R.^{i+1} n_{i+1} +^i P_{C_i} \times^i F_i +^i P_{i+1} \times_{i+1}^i R.^{i+1} f_{i+1}$$

$$\tau_i =^i n_i^{T} .^i Z_i$$

The generalized force at joint (i) is then

$$\mu_i =^i n_i^{T} .^i Z_i + I_{m_i} \ddot{\theta}_i + F_{s_i} sgn(\dot{\theta}_i) + F_{v_i} \dot{\theta}_i + \tau_{v_i} \qquad (2.130)$$

2.8 Matlab Examples

Taking advantage of the preceding discussion, namely the application to the specific manipulator used for demonstration, along with the particularities of *Matlab*, a few functions were built to show how the above presented results could be used to simulate and operate a robot from *Matlab*. The functionality of this collection of functions is extended by the developments presented in chapter's 3 and 4 of this book, which enable the user to command the real robot from the *Matlab* shell.

Several functions were implemented to compute the direct and inverse kinematics, any rotation or transformation matrix, the jacobian (using the method presented here or the differential method presented in [25]), the DLS jacobian, trajectories in the Cartesian or in the joint space, simulate the operation of the robot, etc. The functions developed are related with the robot used for demonstration (ABB IRB1400), i.e., there was no effort to make them compatible with any other type of industrial robot. Consequently, the presented functions were optimized for anthropomorphic robots with a spherical wrist, with the direct and inverse kinematics obtained symbolically using *Matlab* and further optimized.

To demonstrate the functionality of the developed functions, a few examples will be given below.

Jacobian
Functions: jacobian.m and jacobdls.m
Parameters: jacobian (dh, q, type) and jacobdls(dh, q, type) where,
'dh' - *Denavit-Hartenberg* parameters od the robot
'q' - vector or array of vectors containing the joint angles representing a configuration or a sequence of configurations of the robot
'type' - method used to compute the jacobian:
'a' - returns the base jacobian and the *end-effector* jabobian of using differential method presented in [25]
'b' - returns the base jacobian using the same method [25]
'e' - returns the base jacobian using the kinematics developed in this book
'd' - returns the both jacobians using the kinematics developed in this book
'f' - returns the *end-effector* jacobian using the kinematics developed in this book

Figure 2.22 shows the utilization of the above functions to compute the jacobian of the robot for the configuration $q_1 = (0\ 0\ 0\ 0\ 0\ 0)$.

```
» flops(0)
» J=jacobian(dh,q1,'e')

J =

     0  -720  -120     0     0     0
   955     0     0     0     0     0
     0   805   805     0    85     0
     0     0     0     1     0     1
     0    -1    -1     0    -1     0
     1     0     0     0     0     0

» flops

ans =

   188

» flops(0)
» J=jacobian(dh,q1,'b')

J =

   0.0000 -720.0000 -120.0000        0    0.0000        0
 955.0000    0.0000    0.0000        0    0.0000        0
   0.0000  805.0000  805.0000        0   85.0000        0
        0        0        0   1.0000        0   1.0000
   0.0000   -1.0000   -1.0000  -0.0000   -1.0000  -0.0000
   1.0000    0.0000    0.0000  -0.0000    0.0000  -0.0000

» flops

ans =

     3412
```

Figure 2.22 Computing the jacobian: note the reduction of floating point operations when the optimized kinematics is used.

Inverse Kinematics
Function: irb14ink.m
Parameters: irb14ink(dh, t06, quad) where,
'dh' - *Denavit-Hartenberg* parameters of the robot

't06' - Transformation matrix T_6^0 that describes the position/orientation of the terminal element in terms of the base frame
'quad' – indication of the working quadrant. If nothing is given, the routine admits that the working quadrant is equal to the quadrant of θ_1

Figure 2.23 shows the function running applied to a singular configuration with indication of the working quadrant.

```
» qc

qc =

    0.7854    1.0472    0.7854        0        0        0

» t06=irb14mtr(qc',0,0,6)

t06 =

  1.0e+003 *

   -0.0007    0.0007   -0.0002   -0.4906
   -0.0007   -0.0007   -0.0002   -0.4906
   -0.0003    0.0000    0.0010    1.5215
        0         0         0    0.0010

» irb14ink(dh,t06,'q1')
Singular Point -> sin(q5)=0
Resolving Singular Point ...

ans =

   45.0000   45.0000   45.0000    0.7854    0.7854    0.7854
   60.0000   60.0000   60.0000    1.0472    1.0472    1.0472
   45.0000   45.0000   45.0000    0.7854    0.7854    0.7854
        0   -90.0000   90.0000        0   -1.5708    1.5708
        0         0         0        0         0         0
        0    90.0000  -90.0000        0    1.5708   -1.5708
   57.2958   57.2958   57.2958    1.0000    1.0000    1.0000
```

Figure 2.23 Computing the inverse kinematics (initial robot configuration expressed in radians)

2.9 Robot Control Systems

Robot control systems (Figure 2.24) are electronic programmable systems responsible for moving and controlling the robot manipulator, providing also the means to interface with the environment and the necessary mechanisms to interface with regular and advanced users or operators.

In this section, a brief overview of actual industrial robot control systems is presented, pointing out the important factors that must be addressed either by the advanced user (programmer or system integrator) or by the simple operator. Although the discussion is kept general and valid for any robot controller, a particular robot control system (the ABB IRC5 robot controller [26]) will be used for demonstration.

The robot controller has some important tasks it should perform in order to move and control the robot manipulator, provide means for inter-controller and computer communications, enable a sensor interface, and offer the necessary mechanisms and features that allow robot programming, a robot-user interface and program execution.

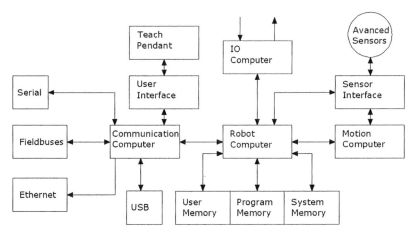

Figure 2.24 Basic architecture of a robot control system

2.9.1 Drive the motors to move the TCP and coordinate the motion for useful work

Motion control involves several different tasks, as already mentioned and resumed in Figure 2.25.

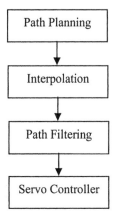

Figure 2.25 Basic tasks involved in motion control

The path planner's basic task is to prepare the robot's path and feed the relevant data to the path interpolator. Moving a robot means specifying an origin position/orientation $\{T_i\}$ and a final position/orientation $\{T_f\}$ of the robot's TCP (*tool center point*). The path interpolator takes the planner data and computes the intermediate points in each interpolation interval, using the specified velocity and acceleration. The outputs of the interpolator are the basic inputs for the servo loops, i.e., they constitute the target points (references) that must be achieved by the servo controllers. The data from the interpolator is filtered by the path filter, before being passed to the servo controllers, in order to provide smoother accelerations/decelerations and keep the motor torques in the range of the servo-motor.

A complete definition of the motion parameters, including velocities and accelerations, is also necessary. Sometimes it is necessary to define intermediate position/orientation points (also called "*via points*") between the initial and final configurations. This procedure will better define the requirements and contribute for the final path. Furthermore, to obtain smooth paths the path planner must be a continuous function, with a continuous first derivative and hopefully also a continuous second derivative [1]. For example, the path generator can be implemented by a 5th order polynomial. The use of a high-order polynomial here is motivated by the fact that a quintic polynomial is needed to be able to specify the position, velocity, and acceleration at the beginning and end of each path segment.

Considering a 5th order polynomial in the form

$$\theta(t) = a_0 + a_1 t + a_2 t^2 + a_3 t^3 + a_4 t^4 + a_5 t^5 \qquad (2.131)$$

with the following constraints

$$\theta_0 = a_0$$
$$\theta_f = a_0 + a_1 t_f + a_2 t_f^2 + a_3 t_f^3 + a_4 t_f^4 + a_5 t_f^5$$
$$\dot{\theta}_0 = a_1$$
$$\dot{\theta}_f = a_1 + 2a_2 t_f + 3a_3 t_f^2 + 4a_4 t_f^3 + 5a_5 t_f^4$$
$$\ddot{\theta}_0 = 2a_2$$
$$\ddot{\theta}_f = 2a_2 + 6a_3 t_f + 12a_4 t_f^2 + 20a_5 t_f^3 \qquad (2.132)$$

Results in a linear system of six equations with six unknowns whose solutions are

$$a_1 = \theta_0$$
$$a_1 = \dot{\theta}_0$$
$$a_2 = \frac{\ddot{\theta}_0}{2}$$
$$a_3 = \frac{20\theta_f - 20\theta_0 - (8\dot{\theta}_f + 12\dot{\theta}_0)t_f - (3\ddot{\theta}_0 - \ddot{\theta}_f)t_f^2}{2t_f^3}$$

$$a_4 = \frac{30\theta_0 - 30\theta_f + (14\dot{\theta}_f + 16\dot{\theta}_0)t_f - (3\ddot{\theta}_0 - 2\ddot{\theta}_f)t_f^2}{2t_f^4}$$

$$a_5 = \frac{12\theta_f - 12\theta_0 - (6\dot{\theta}_f + 6\dot{\theta}_0)t_f - (\ddot{\theta}_0 - \ddot{\theta}_f)t_f^2}{2t_f^5} \qquad (2.133)$$

There are several methods in the literature to compute smooth paths that pass to a given set of "*via points*" [27, 28]. Nevertheless, the function presented above gives a good indication and can be used for that objective, running the function between the intermediate points.

The following *Matlab* functions (Figure 2.26) calculate the robot's trajectory in the joint space using the 5th order polynomial presented above. As already mentioned, with this trajectory planner it is possible to compute the trajectory between two configurations, defining the initial and final velocities and accelerations. The trajectory is represented using a small function that animates the motion of the robot.

Trajectory generation and robot animation
Funtions: irb14trj.m and irb14plt.m
Parameters: [qt, qdt, qddt] = irb14trj(q0, q1, nt, qd0, qd1, qdd0, qdd1) and irb14plt(dh, q, opt, number, azm, elv, vgax, vgay) where,
'q0' – initial position
'q1' – final position
'nt' – number of intermediate points of the trajectory to obtain
'qd0' and 'qd1' – initial and final values of the velocity
'qdd0' and 'qdd1' – initial and final values of the acceleration
'dh' – *Denavit-Hartenberg* parameters of the robot
'q' – matrix holding the computed trajctory
'opt' – type of representation of the motion

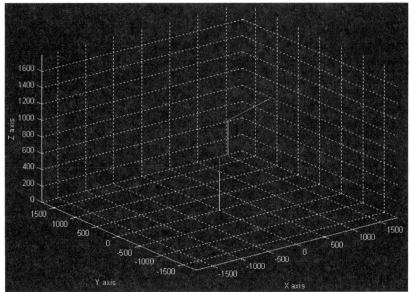

Figure 2.26 Robot's animation using the obtained trajectory

2.10 Servo Control

The servo controllers utilize the data from the path planner and interpolator, properly filtered, to drive the robot manipulator axis. As already mentioned the dynamics of the robot is very complex with a huge number of effects, forces and moments to account for, which puts a considerable challenge to the task of controlling a servo-motor. A detailed and complete description of a servo-controller, namely about the control algorithms and circuitry used, is out of the scope of this book, but a brief overview will be given. Generally, the control loop of an industrial robot joint (or axis) has the components presented in Figure 2.27.

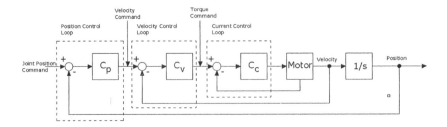

Figure 2.27 Typical robot joint control loop

A brief overview of the AC motors used with industrial robots was already presented, and a typical current control loop was also already sketched in Figure 2.20. Basically, the current control loop implements a PI (proportional and integral) controller [29], having the I component of the controller (Cc) with the objective of eliminating the steady-state error and achieving the best possible control. The velocity control loop is built around the current control loop and also uses a PI controller (Cv).

Finally, around both of the previous controllers there is the position control loop. This controller takes the position commands as input, generates an error signal by subtracting the actual position (obtained from the joint position sensors) from the commanded reference, and generates the control signal using some selected control law (Cp). Typically, the position controller is a simple proportional controller, since the objective is to obtain a good responsive control of the motor position to follow the desired joint command with zero steady-state error and zero overshoot. And that objective is obtained with the combined effect of the position (generally a P controller), velocity (generally a PI controller), and current (generally a PI controller) control loops.

2.11 IO Control

One of the most basic things that a robot control system must do is to implement PLC-like functions. Robots are used in manufacturing cells where digital/analog IO and logic controllers govern the way things happen, namely controlling the systems responsible for material handling, transportation, detection, etc.. To interface with those systems, the robot controller needs to "speak" the same language and act as a logic controller, or at least have the same functionality available. Consequently, the robot controller must be able to:

1. Accommodate digital IO signals with variable and configurable electric levels. The robot must be able to read from digital input lines (with different electric levels) and implement basic logic functions on the obtained data: block reading, logic functions, shifting, counters, timers, edge detection, etc. The robot controller must also be able to act on digital IO outputs changing their state (ON/OFF), applying timed pulses, etc.
2. Accommodate analog IO signals. The robot must be able to read from analog inputs, providing the necessary electronic circuits for multiplexing and analog-to-digital conversion, the mathematical functions to handle the results, and the necessary circuits and digital-to-analog converters to act on analog output signals.
3. Implement IO manipulating functions.

The robot controller programming language must implement advanced mathematical functions, and data structures, that can be used within the robot's program to enable the user to coordinate the robot's motion with IO actions (Figure 2.28), like reading IO information or acting on IO lines (open/close grippers, regulate pressure of pneumatic actuators, regulate the velocity of external motors driven by power inverters or external servo controllers, start/stop equipment, etc.)

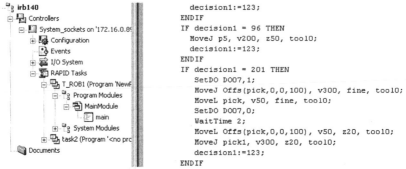

Figure 2.28 Part of a robot program written in RAPID (*ABB Robotics* programming language)

2.12 Communication

Robots are to used in networks with other robots and computers organized into manufacturing cells that also connect to each other constituting manufacturing lines. This type of manufacturing organization corresponds to one of the most recent developments in the area of industrial automation, i.e., the concept of *flexible manufacturing systems* (FMS). These are highly computerized systems composed by several types of equipment, usually connected through a local area network (local network using MAP[13] protocols [30]) under some hierarchical *computer integrated manufacturing* (CIM) structure [31-33]. The available factory (*shop floor*) equipment is organized into *flexible manufacturing cells* (FMCs) with transportation devices connecting the FMCs. In some cases, functionally related FMCs are organized into *flexible manufacturing lines* (FMLs). Each FML may include several FMCs with different or equal basic capabilities. The organization proposed in Figure 2.29 is a hierarchical structure [33,34] where each FMC has its own controller. Therefore, if the manufacturing process is conveniently organized as a FML, then several controllers will exist on the shop floor level, e.g., one controller for each FML. With this setup, an intelligent and distributed job dispatching and awarding may be implemented, taking advantage of the installed industrial network [33,35-37].

---

[13] *Manufacturing automation protocol* (MAP).

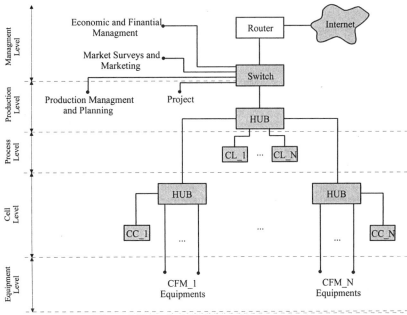

Figure 2.29 Typical CIM hierarchical organization

The best characteristic of an FMC is its flexibility, i.e., its adaptability to new manufacturing requirements that can go from a modified product to a completely new product. The flexibility results from the fact that FMC equipment is programmable and easily reconfigured: that is the case of industrial robot manipulators, mobile robots for parts handling and transportation, *programmable and logic controllers* (PLC), CNC machines, vision systems, conveyors, etc.

Considering the communication between commanding and supervising computers and the robot controllers, and even the communication between robot controllers itself, it is usually supported through a TCP/IP *Ethernet* based network. The functions associated with this type of communication include the exchange of files and programs, the execution of remote operations like backup and system maintenance, etc. In many advanced applications, this network is also used to command and supervise each manufacturing cell operation, with several levels of functionality depending on the type of access: operator access, supervisor access, or information retrieval access from the production planning levels of the network. These types of advanced features will be extensively explored in this book.

Many manufacturers offer robot services through this type of network to support these advanced applications, in the form of RPC servers [38], TCP/IP socket servers [26], or UDP datagram servers [39]. These servers and associated services can be used by the system developer/integrator to provide functionality to the user through the application.

Furthermore, the communication links between the controller and the manufacturing cell can be as follows:

1.   Computer network – to interface with commanding and supervising computers, from several levels of the network
2.   Fieldbuses – to interface with other robot controllers, but also with PLCs and other cell equipment commanded by programmable controllers. The most common options are *DeviceNet*, *ProfiBus*, *Ethernet IP*, etc. Several robot controllers also use a fieldbus network (*CAN* or *DeviceNet*, for example) to connect some of its internal components (the drive boards to the main computer, etc.)
3.   Serial IO – to interface with sensors, or with several types of IO equipment or process equipment like welding power sources, to interface locally with a computer or laptop using a point-to-point occasional connection, and so on

2.13 Sensor Interface

Interfacing advanced sensors is a fundamental aspect of any robot control system. In fact, to successfully perform several actual industrial tasks, the robots need special sensors that could be used to help them get the relevant information and use it efficiently through the process. Many of these sensors require high-performance, non-perturbed communication links, and/or need to interface directly to the path planners and motion controllers so that the robot can be guided and instructed in real-time. Consequently, the robot controllers should provide special interfaces for these types of sensors, at least for the most common ones, which can be programmed and explored by the advanced user.

2.13.1 Interfacing Laser 3D Sensor for Seam Tracking

Good examples are the laser sensors used in robotic welding for seam finding and tracking during the welding operation. These types of sensors provide signals (analog or through high-speed digital interfaces) that can be used to guide the robot during the welding operation. These sensors work in a simple way, based on the principle of laser triangulation. A low power laser source is used to generate a laser beam that is projected onto the surface of the joint to weld. The reflected light is picked up by a lens that feeds the imaging system, composed usually of a CCD or CMOS sensor. The laser-reflected signals are extracted using filters and image processing software, which is a simple task since the laser signal has a very precise wave length and power (Figure 2.30).

In fact, these laser cameras and related processing hardware and software, with some customization to the selected application, are useful for evaluating most of the geometric parameters other than the mentioned joint detection and seam tracking features. Since they are available with powerful APIs for general use, with standard interfaces for robot controllers and current computer hardware, these types of sensors constitute a powerful tool for robotic welding.

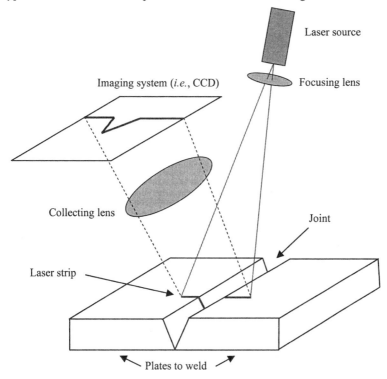

Figure 2. 30 Explanation of the laser vision principle

Basically, the outputs obtained from these sensors are position accommodations, or position corrections, that should be sent to the robot controller to adapt the current motion. They can also monitor certain variables and provide the means to generate interrupts in the robot controller in order to respond to significant variable changes. For example, the seam volume or the welding gap can be monitored by this sensor. When changes are detected, the corresponding events can be used to trigger an internal interrupt that will adapt the welding parameters (voltage, wire feed and velocity) accordingly. For example, the following would be the procedure to adapt the welding parameters in function of the measured welding gap:

Variables
Matrix Numeric Adapted_voltage = {1, 1.1, 1.2, 1.4, 1.6, 2, 2.2, ...};
Matrix Numeric Adapted_wire_feed = {2, 2.2, 2.4, 2.6, 2.8, 3, 3.2, ...};
Matrix Numeric Adapted_velocity = {10, 12, 14, 16, 18, 20, 22, ...};
Numeric gap_value;

Numeric index;

Program
Set Interrupt 1 when gap_value changes;
Start Welding, tracking;
When target point achieved
Stop welding, tracking;
EndWhen
EndProgram

Interrup Service Routine
index = scale(gap_value);
voltage = adapted_voltage(index);
wire_feed = adapted_wire_feed(index);
velocity = adapted_velocity(index);
refresh welding parameters;
EndRoutine

The position of the sensor can also be read and used to accommodate the position references sent to the motion controller, guiding in this way the robot's motion.

The next example shows how to interface other type of intelligent sensors for which there is no special interface at the robot controller.

2.13.2 Interfacing a Force/Torque Sensor

As already mentioned, robot manipulators are good examples of equipment for *flexible manufacturing systems*, due to their flexibility. In fact, flexibility is the major reason for robot utilization and popularity in actual manufacturing plants. In this framework, the majority of the robot's tasks require contact with the surrounding environment, i.e., in the process of fulfilling the task, the robot tool interacts physically with the working objects and surfaces. That interaction generates contact forces that should be controlled in a way to finish the task correctly, not damaging the robot tools and working objects. Those contact forces depend on the stiffness of the tool and working objects/surfaces and should be properly controlled. The option for a particular control technique depends on identifying if [40]:

1. The contact forces should be controlled to achieve task success, but are sufficient to keep them inside some safety domain: *passive force control* [40].
2. The contact forces should be controlled because they contribute directly to the success of the task: *active force control* [40-53].

In the first case, contact forces are an undesirable effect of the task and it is generally sufficient to keep them inside some safety domain. They are not necessary for the task, so usually the strategy is adding flexibility to the *end-effector* with the object of damping all the possible impacts and increasing the tolerance to positioning errors, complemented with detailed and careful planning of flying trajectories and object approach. There are many solutions in the market to add flexibility to the *end-effector*, and in fact this is currently the standard approach in industry.

In the second case, the contact forces are necessary to finish the task correctly, i.e., controlling the contact forces to make them assume some particular value or, more generally, to follow some force profile.

For industrial robotics applications, force/torque sensors are usually placed near the working tool, generally in the manipulator wrist. This means that the sensor must be reasonably small, built in several sizes to adapt to different robot bolt patterns and load capacities, and mechanically resistant. Considering these restrictions, it is easy to understand why measuring the strain imposed on a selected strain gauge material, just by reading the voltage across the resistance of the material, is still the most used sensing technique.

There are several ways and materials to design sensing gauges, metal wire, metal-foil and semiconductor gauges being the most common. From those, the metal-foil gauges show some interesting features. The strain induced change in resistance is due to length and sectional area changes as well as a small piezo-resistive effect. With the developments in etching processes, metal-foil gauges became a very interesting possibility. They are manufactured in very thin foils (less than 10 μm), with sizes down to 200 μm, etched by photographic methods. Consequently, there are virtually no limits to the variety of possible geometries. This gives greater flexibility to design geometries, but also to the type of stressing at the surface of the elastic material component where the gauge will be attached. Metal-foil gauges have very high linearity, with very low transverse sensibility (less than 0.3%), and great dynamic range. Also, their thermal characteristics are better than their semiconductor and metal-wire counterparts. All these arguments explain why metal-foil gauges are ideal for force/torque sensing elements. Force/torque sensors manufactured by JR3 (the ones we use in this book) use metal-foil gauges bound to elastic rings as sensing elements, which explain their superior behavior. Figure 2.31 shows the composition of these sensors.

The sensing part. It is composed of elastic rings at the outer perimeter between the mounting plates. The monolithic design eliminates hysterisis that would occur from slippage at highly stressed internal joints. The use of elastic rings produces a very stiff device, resulting in minimal deflection under load and better performance at higher frequencies. The rings and their strain gauges are positioned so that the local strain measures can be used to deduce the forces and moments, in all cartesian directions (X, Y, Z), passing through the sensor. The internal cavity between the mounting plates contains the front-end electronics where signals are amplified, digitized, and transmitted to the host receiver board. If the amplification and digitization electronics are inside the sensor, preferable for noisy or industrial environments, there is no analog signal being transmitted and high sampling rates can be achieved (8Khz).

Table 2.3 Functions available in the MATJR3PCI *Matlab* Mex file

Functions	Brief description
init_jr3	This function opens a handle to the JR3PCI driver, checks memory, and downloads DSP code to the board.
read	Reads from a receiver board memory address.
write	Writes to a receiver board memory address.
system_warnings	Reads system saturation warnings (board memory address WARNINGS).
system_errors	Reads system errors (board memory address ERRORS).
command	Commands JR3 receiver board.
get_threshold_status	Gets the value of the threshold bits (board address THRESHOLD).
reset_threshold	Resets the threshold bits.
read_ftdata	Reads force/torque data from receiver board.
set_transforms	Sets a new transformation definition.
use_transforms	Selects the transformation to use.
read_offsets	Read offsets in use.
set_offsets	Sets actual offsets, using the current offset index.
change_offset_num	Changes actual offset index (num).
reset_offsets	Sets actual offsets to the current values read from FILTER_2.
use_offset	Changes actual offsets to the one defined.
peak_data	Sets address to watch for peaks.
peak_data_reset	Sets address to watch for peaks and resets internal values to current data.
read_peaks	Reads current peak values.
bit_set	Sets bits on defined bit-map.
set_full_scales	Sets JR3 Full_Scales.
get_full_scales	Reads actual full_scales.
get_recommended_full_scales	Reads recommended full_scales.
sensor_info	Reads information from the sensor and from the receiver board. Use this function to test your setup.

Note: all these functions address a specific sensor, even if a multi-channel board is used.

DSP receiver board. Based on the same basic architecture, several interfaces can be used. If the issue is high access rates, then fast IO buses must be used and a shared memory mechanism must be implemented to exchange data and program the sensor. JR3 offers several interface buses like VME, PCI (up to four channels per board), CPCI (also up to four channels) and ISA. The receiver boards are basically DSP boards that implement digital filters and dispose sensor information to users. Also they parameterize readings (offsets, full scales, geometrical transformations, etc.) and implement a few interesting functions such as maximum and minimum values (peaks) and, warning and error bits, etc. A full description of these functions can be found in [54], and a brief summary can be found in Table 2.3.

Interface software and drivers. For *Win32*-based operating systems, we developed a complete set of tools that can be used to build applications using force/torque sensors. These tools include kernel drivers designed for *Win32* operating systems, i.e., *Windows*. Basically, when we want to use some kind of equipment from a computer we need to write code and define data structures to handle all its functionality. We can then pack the software into libraries, which are not easy to distribute being language dependant, or build a software control using one of the several standard languages available. Having in mind that force/torque sensors can be used by persons with different programming capabilities, and from several types of programming languages and environments, the collection of functions that access the sensor capabilities are offered in several packages: C++ Library, *ActiveX* software component, *Matlab* toolbox and *LabView Virtual Instruments* [55].

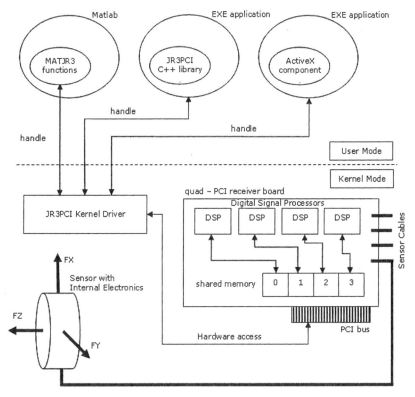

Figure 2.31 Force/torque sensor overview (using PCI receiver board)

With this organization, the sensor works like a server, offering a collection of services to the advanced user, who can use the available programming tools cited above to tailor the sensor behavior. The next section demonstrates the sensor capabilities using the popular application *Matlab*.

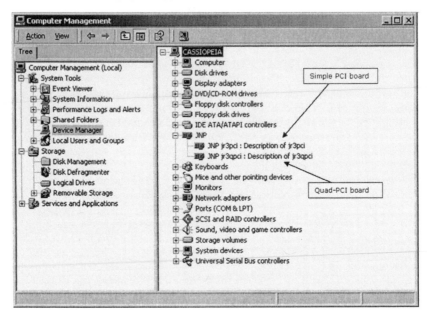

Figure 2.32 Boards reported by Windows device manager

2.13.2.1 Using a Force/Torque Sensor

There are several applications of force/torque sensors, but generally a user just wants to install the sensor on his computer (after installing the sensing part on the mechanical system he is using), and then be able to parameterize it and get the sensor readings at selected rate from within the selected environment he chose to use. The basic software [54] was prepared to be used with virtually any application or programming language under *Win32* operating systems, by any type of user: from computer experts to regular users. Here we use two different environments to explore the sensor capabilities. In this section, *Matlab* is used. *Matlab* is a widely used software environment for research and teaching applications on robotics and automation, mainly because it is a powerful linear algebra tool, with a very good collection of toolboxes that extend its basic functionality, and because it is an interactive open environment. So, it is really a good environment to demonstrate how to use this type of intelligent sensor.

From all the available receiver board models, the quad-PCI receiver model was used. This board is capable of handling four force/torque sensors at the same time on a single PCI slot. It will be used step-by-step.

After having the board installed and correctly reported by the operating system (Figure 2.32), with sensor cables attached, the user is ready to start using the sensor. The first thing to do is open a handle to the sensor receiver board, check if the board is OK, and download the DSP code to the receiver's board program memory.

The command is

>> matjr3pci('init_jr3',vendor_ID, device_ID, n_board, n_proc,download);

where *vendor_ID* and *device_ID* are the PCI ID's of the selected board, *n_board* is the board number (there can be several in the PCI bus), *n_proc* is the number of DSP units in the board, and *download* specifies if the user wants to download (1) the DSP code to the program memory or not (0). Nevertheless, DSP code must be downloaded once after each computer power-up, but after that the command can be used simply to open a handle to the board. The command returns zero if successful, or an error code [45]. Consequently, to a quad-PCI board, the command with DSP code download should be:

>> matjr3pci('init_jr3', 0x1762, 0x3114, 0, 4, 1);

or without download:

>> matjr3pci('init_jr3', 0x1762, 0x3114, 0, 4, 0);

If the return value is zero (0) then the user can start using the sensor, otherwise the user must solve the problem reported by the software (error code).

The first command could be a query to the system to find out what sensor is attached to each channel. The command is

>> matjr3pci('sensor_info', 2);

to get information about the force/torque sensor handled by DSP number 2. The returned information includes model and serial numbers, software version, number of ADC bits, etc.

To read offsets from the force/torque sensor handled by DSP number zero (remember we are using a board with 4 DSP: numbered from 0 to 3),

>> offset_matrix = matjr3pci('read_offsets', 0);

To set offsets of the force/torque sensor handled by DSP number 2,

>> matjr3pci('set_offsets',matrix_off, 2);

where *matrix_off* is a matrix with the offset values.

To reset offsets,

```
>> matjr3pci('reset_offsets', n_dsp);
```

where n_dsp is the DSP number. With this function, the offsets are zeroed using the actual values reported by FILTER_2 [56].

The offsets are stored in the memory available for each DSP. It is possible to store 16 independent tables of offsets for each DSP. Consequently, before any of the previous operations, the user should define the table currently in use. If the definition is not performed, all operations are referent to the actual table. To set a table for offset reading the command is,

```
>> matjr3pci('change_offset_num, 12, 1);
```

to specify that all subsequent offset operations for the sensor handled by DSP number 1 are to be addressed to Table 12. Table 12 is also used on any subsequent force/torque reading for that sensor.

To specify a table for actual force/torque readings the command is,

```
>> matjr3pci('use_offset, 10, 2);
```

where table 10 was selected for sensor handled by DSP number 2.

Another important operation on this type of sensor is setting the full-scales to properly scale the readings. This operation is similar to the operations of setting and reading offsets, so it will not be mentioned explicitly.

Each DSP has an address space [56]. To read, write, and issue commands relative to those address spaces the user should use the *read, write,* and *command_jr3* commands. For example, to read the serial number (address 0x00f8 of each DSP address space) of the force/torque sensor attached to DSP number 2 the command is,

```
>> serial_2 = matjr3pci('read_jr3', 248, 2);
```

Finally, to read data from any sensor the command is,

```
>> ft_data = matjr3pci('read_ftdata', n_filter, n_dsp);
```

where n_filter is the filter number (from 0 to 6, where 0 means unfiltered data), and n_dsp is the DSP number.

The collection of functions available from this *Matlab* toolbox can be found in [54] and the correspondent functions of the C++ library or *ActiveX* control can be found in [57]. The same basic function prototypes have been kept between all the software packages, which makes the above *Matlab* demonstration a good way to show how the other packages work (C++ library, *ActiveX* control, etc).

This example demonstrates how to interface an intelligent sensor to a computer. If the same facilities were available from the robot controller, then it would be equally easy to make the interface available directly from the controller, enabling in this way the programmer to directly use its readings to influence the robot's motion. Nevertheless, with most of the commercial robot controllers, this type of advanced access is not available or isn't accessible. Consequently, these types of sensors must be used form personal computers feeding the data to the robot using the available communication channels. This type of indirect approach slows down the possible performance, but it's an alternative way to implement the interface to the force/torque sensors.

2.14 Programming and Program Execution

Robot controllers should provide a programming language and a library of functions to enable users to explore the functionalities of the robot and of the robot's controller. Most of the manufacturers offer advanced *PASCAL-like* structured programming languages, including a language interpreter within the controller. Consequently, users can write code using any ASCII editor, download it to the controller, and run it immediately without the need for any type of file conversion. Those programming environments also offer simple debugging tools that make the process of developing software easy.

The most advanced manufacturers also offer online and offline PC-based programming tools, which enable users to develop code directly in the controller (online) using a remote PC. Alternatively, the code can be developed offline and downloaded to the controller when ready.

The Teach Pendant Unit (TPU) can also be used to program and parameterize the system. These devices are basically computer units running a local operating system (*Windows CE*, for example) that offer to several types of users the possibility to program, parameterize, and operate the robot manipulator.

The actual robot controllers are also multitasking systems, which enable the user to develop and run multiple tasks simultaneously. This allows new levels of functionality, offering new possibilities to the system developer. Using the available and common inter-task communication mechanisms, along with the ability to regulate task priorities (percentage of CPU time), it's possible to set up applications to handle all the challenges posed by the industrial manufacturing cells.

2.15 User Interface

The user interface is basically defined by the system developer, because there are a lot of possibilities. The developer can use the available communication links and the robot controller's remote servers to set up a PC interface to command and monitor the robot operation (see for example Figures 1.20 and 1.21). Alternatively, he can use the controller TPU to design the user interface. Since most of the current teach pendants are advanced computers, running powerful operating systems, the possibilities for developing advanced interfaces are enormous and flexible.

For example, the TPU that comes with the new ABB IRC5 controller [26] is a *Windows CE* system (Figure 2.33), equivalent to any portable CE based consumer device, which can be programmed remotely from a PC using common programming tools like the *Microsoft Visual Studio .NET* programming suite.

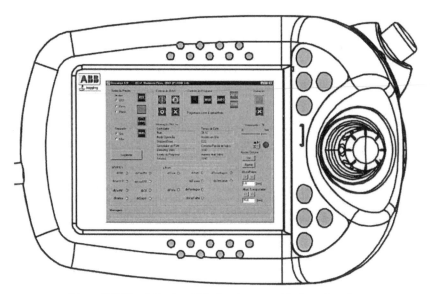

Figure 2.33 Teach Pendant Unit showing a graphical user interface

This book explores several examples that use a remote PC to implement the user interface, examples that use mainly the TPU, and examples that use both possibilities. The idea is to demonstrate that the possibilities are there and that it's up to the system developer to pick the best options for the specific application he's building.

2.16 References

[1] Craig, J.J., "Introduction to Robotics, Mechanics and Control", 2ª Edition, Addison-Wesley, 1989.

[2] Sciavicco, L., and Siciliano, B., "Modeling and Control of Robot Manipulators"-2nd Edition, McGraw-Hill, 1996.

[3] De Wit, C.C., Siciliano, B., Bastin B., "Theory of Robot Control", Springer-Verlag, London, 1996.

[4] Hollerbach, J.M., "A Recursive Lagrangian Formulation of Manipulator Dynamics and a Comparative Study of Dynamics Formulation Complexity", IEEE Transactions on Systems, Man and Cybernetics, Novembro de 1980.

[5] Pieper, D.L., "The Kinematics of Manipulators Under Computer Control", memo. AIM72, Stanford Artificial Intelligence Laboratory, 1968.

[6] Symon, K.R., "Mechanics", 3ª Edition, Addison-Wesley, 1971.

[7] Fu, K., Gonzalez, R., Lee, C.S.G., "Robotics: Control, Sensing, Vision and Intelligence", McGraw-Hill, 1987.

[8] Klema, V.C., Laub, A.J., "The Singular Value Decomposition: Its Computation and Some Applications", IEEE Transactions on Automatic Control, Vol. AC-25, N° 2, April 1980.

[9] Chiaverini, S., Siciliano, B., Egeland, O., "Review of the Damped Least-Squares Inverse Kinematics with Experiments on an Industrial Robot Manipulator", IEEE Transactions on Control Systems Technology, Vol.2, N°2, June 1994.

[10] Golub, G.H., Van Loan, C.F., "Matrix Computations", The Johns Hopkins University Press, Baltimore, Maryland, 1983.

[11] Wampler, C.W., Leifer, L.J., "Applications of Damped Least-Squares Methods to Resolved-Rate and Resolved-Acceleration Control of Manipulators", Journal of Dynamical Systems, Measurement and Control, Vol.110, January of 1988.

[12] Golub, G.H., Klema, V.C., Stewart, G.W., "Rank Degeneracy and Least Squares Problems", Department of Computer Science, Stanford University, Technical Report STAN-CS-76-559, August 1976.

[13] Maciejewski, A.A., Klein, C.A., "Numerical Filtering for the Operation of Robotic Manipulators through Kinematically Singular Configurations", Journal Robotics Systems, Vol.5, 1988.

[14] Chiaverini, S., "Estimate of the Two Smallest Singular Values of the Jacobian Matrix: Application to Damped Least-Squares Inverse Kinematics", Journal of Robotic Systems, Vol.10, N°8, 1993.

[15] Luh, J.Y.S., Walker, M.W., Paul, R.P.C., "Resolved-Acceleration Control of Mechanical Manipulators", IEEE Transactions on Automatic Control, Vol.AC-25, 1980.

[16] Dote, Y., "Servo Motor and Motion Control using Digital Signal Processors", Texas Instruments and Prentice-Hall, 1990.

[17] Herdershot Jr., J.R., and Miller, T.J.E., "Design of Brushless Permanent-Magnet Motors", Magna Physics Publishing and Clarendon Press, Oxford, 1995.

[18] Crowder, R.M., "Electric Drives and their Controls", Clarendon Press, Oxford, 1995.

[19] Tamagawa Seiki Co. Ltd, "SmartSyn Brushless Resolvers, General Catalog", Tamagawa Seiki, Japan, 2005.

[20] ABB Robotics, "S4-IRB1400 Product Manual" - M94A, ABB Flexible Automation, 1994.

[21] Hanselman, D.C., "Techniques for Improving Resolver-to-Digital Conversion Accuracy", IEEE Transactions on Industrial Electronics, Vol.38, No. 6, December 1991.

[22] Boyes, G., "Synchro and Resolver Conversion", Analog Devices Inc. (Norwood, MA), 1980.

[23] Analog Devices, "AD2S80A Resolver to Digital Converter - Data Sheet", Data Conversion Manual, 1995.

[24] Goldstein, H., "Classical Mechanics", 2ª Edição, Adison-Wesley, 1980.

[25] Paul, Shimano and Mayer, "Differential Kinematic Control for Simple Manipulators", IEEE Trans. SMC Vol.11, n.6 Junho de 1981.

[26] ABB Robotics, "IRC5 documentation CD", ABB Robotics, 2005

[27] Deboor, C., "A Practical Guide to Splines", Springer, New-York, 1979

[28] Rogers, D., Adams, J.A., "Mathematical Elements for Computer Graphics", McGraw-Hill, 1976.

[29] Ogata, K., "Modern Control Engineering" Prentice-Hall Inc., 1970.

[30] Halsall F., "Data Communications, Computer Networks and Open Systems", Third Edition, Addison-Wesley, 1992.

[31] Kusiak A., "Modelling and Design of Flexible Manufacturing Systems", Elsevier Science Publishers, 1986.

[32] Ou-Yang C. and Lin JS., "The Development of a Hybrid Hierarchical/ Heterarchical Shop Floor Control System Applying Bidding Method in Job Dispatching", Robotics and Computer-Integrated Manufacturing, 1998;14(3):199-217.

[33] Waldner JB., "CIM, Principles of Computer Integrated Manufacturing", John Wiley & Sons, 1992.

[34] Liang GR., "A Hybrid Model of Hierarchical Control Architecture in Automated Manufacturing Systems", in Advances in Factories of the Future, CIM and Robotics, Elsevier Science Publishers, 1993:277-286.

[35] Baker AD., "Complete Manufacturing Control Using a Contract Net: A Simulation Study", Proceedings of the IEEE International Conference on Computer Integrated Manufacturing, 1988: 100-9.

[36] Shaw M., "A Distributed Scheduling Method for Computer Integrated Manufacturing; the use of Local Area Networks in Cellular Systems", International Journal on Production Research, 1987;25(9):1285-1303.

[37] Zhang Y., Kameda H. and Shimizu K., "Adaptive Bidding Load Balance Algorithms in Heterogeneous Distributed Systems", Proceedings of the IEEE Second International Workshop on Modeling, Analysis and Simulation of Computer and Telecommunication Systems, 1994:250-254.

[38] RAP, Service Protocol Definition, ABB Flexible Automation, 1996.

[39] Remote Connection Manual for the NX100 Controller, Motoman Robotics, 2005

[40] Siciliano B., Villani L., *Robot Force Control*, Kluwer Academic Publishers International Series in Engineering and Computer Science, Boston, MA, 1999

[41] Pires, JN, and Sá da Costa JMG, "A Real Time System for Position/Force Control of Mechanical Manipulators", Proceedings of the 7th International Machine Design Conference, Ankara, Turkey, 1996.

[42] De Schutter, J., and Van Brussel H., "Compliant Robot Motion I. A Formalism for Specifying Compliant Motion Tasks", The International Journal of Robotics Research, August de 1988.

[43] De Schutter, J. and Van Brussel, H., "Compliant Robot Motion II. A Control Approach Based on External Control Loops", The International Journal of Robotics Research, August, 1988.

[44] Craig JJ, and M.H Raibert MH, "A Systematic Method of Hybrid Position/Force Control of a Manipulator", IEEE Computer Software Applications Conference, November, 1979.

[45] Nilsson K., "Industrial Robot Programming", Ph.D. Thesis, Department of Automatic Control, Lund Institute of Technology, May of 1996.

[46] Hogan N., "Impedance Control: An Approach to Manipulation: Part I-Theory, Part II-Implementation, Part III-Applications", ASME Journal of Dynamic Systems, Measurement, and Control", March, 1985.

[47] Khatib O., "A unified Approach for Motion and Force Control of Robotic Manipulators: The Operational Space Formulation", IEEE Journal of Robotics and Automation, February 1987.

[48] Volpe R. and Khosla P., "A theorical and Experimantal Investigation of Explicit Force Control Strategies for Manipulators", IEEE Transactions on Automatic Control, November, 1993.

[49] Volpe R. and Khosla P., "An Analysis of Manipulator Force Control Strategies Applied to an Experimentally Derived Model", IEEE/RSJ International Conference on Intelligent Robots and Systems, Raleigh, July, 1992.

[50] Volpe R. and Khosla P., "Computational Considerations in the Implementation of Force Control Strategies", Journal of Intelligent and Robotic Systems, 9-1994

[51] Volpe R. and Khosla P., "On the Equivalence of Second Order Impedance Control and Proportional Gain Explicit Force Control", to appear in The International Journal of Robotics Research, 1994.

[52] Siciliano B., "Parallel Force/Position Control of Robot Manipulators", Proceedings of the 7th International Symposium of Robotics Research, Springer-Verlag, London, UK, 1996:79-89.

[53] Chiaverini, S., "Force/Position Regulation of Compliant Robot Manipulators", IEEE Transactions on Automatic Control, Março de 1994.

[54] Pires, JN, "MATJR3PCI", Users Manual of the JR3PCI Matlab Toolbox, http://robotics.dem.uc.pt/norberto/jr3pci/, 2001.

[55] Pires, JN, "Using Matlab to Interface Industrial Robotic & Automation Equipment", IEEE Robotics and Automation Magazine, September 2000.

[56] JR3 Force/Torque Sensor Users Manual, JR3 Inc. Woodland, California, 2001.

[57] JR3 PCI Web Site, http://robotics.dem.uc.pt/norberto/jr3pci/, 2001.

3

Software Interfaces

3.1 Introduction

This chapter explains the basics of remote procedure calling using robot manipulators and industrial automation systems in general. The underlying idea here is to demonstrate how to set up and explore a basic facility for robot cell commanding and supervision operations, using the available network services. Consequently, a client-server model is adopted where the robot acts like a server exposing to the remote clients its remote services.

The basic idea is simple. For each equipment we need to design and build a server (if it is not yet available) to expose the equipment functionality as remote services. The technology to build the server is highly dependent on the equipment resources and computing facilities, but if possible some kind of RPC (*remote procedure calls*) [1,2] mechanism should be used. Software controls that explore these services should then be available as basic tools to develop remote and distributed applications using the selected equipment.

The OSI (*open systems interconnection*) reference model [1,2] defines the seven basic levels of network communications. The OSI seven layers can be summarized as follows (Figure 3.1):

1. **Physical layer** - Provides electrical, functional, and procedural characteristics to activate, maintain, and deactivate physical links that transparently send the bit stream
2. **Data link layer** - Provides functional and procedural means to transfer data between network entities and eventually correct transmission errors. It also provides mechanisms for activation, maintenance, and deactivation of data link connections, grouping of bits into characters and message frames,

character and frame synchronization, error control, media access control, and flow control

3. **Network layer** - Provides independence from data transfer technology and relaying and routing considerations; masks peculiarities of data transfer media from higher layers and provides switching and routing functions to establish, maintain, and terminate network layer connections and transfer data between users

4. **Transport layer** - Provides transparent transfer of data between systems, relieving upper layers from concern with providing reliable and cost effective data transfer; provides also end-to-end control and information interchange with the quality of service needed by the application program; first true end-to-end layer

5. **Session layer** - Provides mechanisms for organizing and structuring dialogues between application processes; these mechanisms allow for two-way simultaneous or two-way alternate operation, establishment of major and minor synchronization points, and techniques for structuring data exchanges

6. **Presentation layer** - Provides independence to application processes from differences in data representation, i.e., in syntax; syntax selection and conversion provided by allowing the user to select a "presentation context" with conversion between alternative contexts

7. **Application layer** – This layer is dedicated to the requirements of application. Consequently, application processes use the service elements provided by the application layer. The elements include library routines that perform inter-process communication, provide common procedures for constructing application protocols and for accessing the services provided by servers that reside on the network

The user/programmer selects the remote procedure calling mechanism to be used with the application. Ideally, the libraries used should isolate the user from the transport selected, hiding the tricky details about how to handle the communication flow.

This chapter considers the various ways to achieve client-server communication, with the objective of commanding remote execution of selected functions. The final objective is to achieve semi-autonomous systems, i.e., highly automated systems that require only minor operator intervention. In many industries, production is closed tracked in many parts of the manufacturing cycle, which is composed by several in-line manufacturing systems that perform the operations necessary to transform the raw materials into a final product. In many cases, if properly designed, those individual manufacturing systems require simple parameterization to execute their tasks. If that parameterization can be commanded remotely by automatic means from where it is available, then the system becomes almost autonomous in the sense that operator intervention is reduced at a minimum and essentially needed only for error and maintenance situations. A system like this will improve efficiency and agility, since it is less dependent on human operators. Also, since those systems are built under distributed frameworks, based on client-server software architectures that require a collection of functions to implement the system functionality, it is easier to change production by adjusting parameterization (a software task now), which also contributes to agility. Furthermore, since all information about each item produced is available in the manufacturing tracking software, it is logical to use it to command some of the shop floor manufacturing systems, namely the ones that require simple parameterization to work properly. This procedure would take advantage of the available information and computing infrastructure, avoiding unnecessary operator interfaces to command the system. Also, further potential gains in flexibility and productivity are evident.

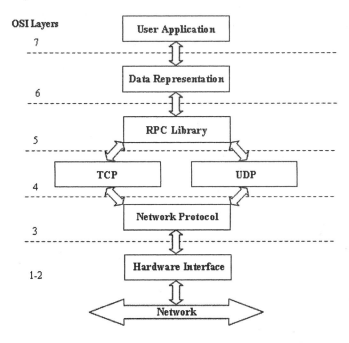

Figure 3.1 OSI reference model, with reference to an RPC library (used in this book)

3.2 Low Level Interfaces

3.2.1 IO Digital Signals

Probably the simplest way to exchange information between two machines, the first acting as *client* and the other as *server*, is by using IO digital signals. Basically, the *client* and the *server* can "agree" to exchange information using a predefined number of IO digital lines and a simple messaging protocol.

Let's illustrate this possibility with an example. Consider the setup represented in Figure 3.2, composed of a robot manipulator equipped with a vacuum suction cup and four fixed pick-place positions defined over a working table.

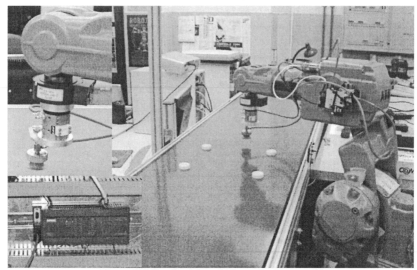

Figure 3.2 Simple pick-and-place robotic example

The user should be able to control the robot from a personal computer (PC), commanding it to pick or place a working piece on any of the available four positions. The user should also be able to start the robot, send it to the "*home position*" and get basic monitoring information.

The commands needed for this application are:

Commands	Parameters
Pick piece from position	P1 to P4
Place piece at position	P1 to P4
Program RUN/STOP	...
Motor ON/OFF	...
Go home	...
Start Vacuum	...
Release Vacuum	...
Get Robot Status	...
Acknowledge Error	...

Therefore, considering all the possibilities there are seventeen different commands that require at least five bits (signals). Furthermore, to include the system commands "*Motor ON*", "*Motor OFF*", "*Program RUN*", and "*Program STOP*" four new digital input signals are needed (defined in the robot controller as

SYSTEM INPUTS). These system commands may be necessary for systems that don't support multitasking, and consequently require systems inputs to implement those actions; we plan to implement the server routine as a semistatic independent task, i.e., a task that runs when the system is in automatic mode. Other systems may require to have those commands associated with independent IO lines. For generality we admit here both scenarios. The synchronization signal "*command ready*" is also needed to signal valid commands.

To add a simple handshaking mechanism to be used to get robot status information (like *busy, ready,* and *error* status information), and system and program state information, another six digital output signals are needed. Consequently, the following IO digital signals should be used:

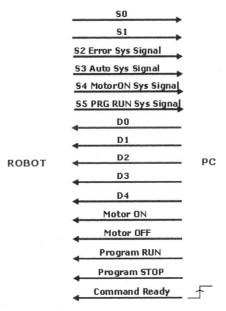

i.e., six (6) robot digital outputs for robot status information and ten (10) robot digital inputs for system command, data communication, and command validation. Consequently, Table 3.1 lists the commands identified for robot command and supervision.

Table 3.1 Commands adopted for this example – PLC side

Command	Value of D0-D4 (Hex)
Pick from P1	01
Pick from P2	02
Pick from P3	03
Pick from P4	04
Place at P1	05
Place at P2	06

Place at P3	07
Place at P4	08
Go home	09
Start Vacuum	0A
Stop Vacuum	0B
Acknowledge Error	0C
Motor ON	0D
Motor OFF	0E
Program RUN	0F
Program STOP	10
Get robot status	1F

The following procedure should be used to run the presented setup:

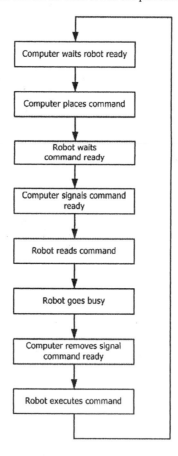

with the following exceptions:

1. The robot only accepts commands when in *automatic mode*. In *manual mode* or *error state* the *robot ready* signal is never activated.
2. When in *manual mode,* the system always returns the *offline state* status.
3. On an error situation, the system returns *error state* status and requires the user to issue a *release error* command.

A simple IO board installed on the PC can be used to support the implementation of the ROBOT – PC interface. Nevertheless, in this example, an industrial PLC was used to implement the IO interface with the robot controller, being the communication between the commanding PC and the PLC done through a serial link (RS232C) – see Figure 3.3. The setup (Figure 3.2) is composed of an industrial PLC (*Siemens S7-200 CPU15*) [2], a personal computer running *Windows XP* and an industrial robot manipulator (ABB IRB 140 equipped with the IRC5 robot controller).

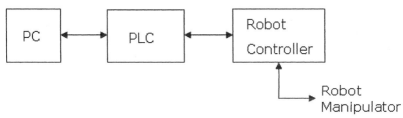

Figure 3.3 Main components of the system: PC (user interface), PLC (Io interface), robot controller, and manipulator

The PLC was designed to operate as a server, offering IO services to the remote computer. Basically, the PLC waits for remote commands, processes them, and returns the status of all the IO signals. The commands have the following format:

$$CMD\ PAR\_1\ PAR\_2\ ...\ PAR\_N$$

where *CMD* is a code that identifies the command (Table 3.2), and *PAR_1* to *PAR_N* are parameters associated with each command.

Table 3.2 Commands adopted for this example – PC side

Command	Code (decimal)	Parameters
Pick	200	1 to 4
Place	201	1 to 4
Go home	202	...
Start Vacuum	203	...
Stop Vacuum	204	...
Acknowledge Error	205	...
Motor ON	206	...
Motor OFF	207	...

Program RUN	208	...
Program STOP	209	...
Get robot status	500	...

In the following few sections, the developed robot software, the PLC server software, and the PC commanding software will be presented and explained.

3.2.1.1 Robot Controller Software
In simple terms, the robot software executes the commands defined for the application in Tables 3.1 and 3.2, following the protocol sequence specified above. Consequently, the code has the basic structure depicted in Figure 3.4 where the RAPID programming language (from *ABB Robotics*) was used. For practical reasons the software presented in Figure 3.4 shows only the basic structure of three types of services: *Pick/Place P1*, *Go Home,* and *Start/Stop Vacuum*. It is assumed here that the robot server routine can run as an independent task, which requires a multitasking robot controller.

```
MODULE server_sock
VAR Declaration Here
...
PROC main()
WHILE TRUE DO
  SetDO s0,1;
  WaitUntil cmd_rdy = 1;
  WaitUntil (command > 0 and command < 15);
  SetDO s0 = 0;
  WaitUntil command_ready = 0;
  TEST command
  CASE 1: ──────────────────────────────►  Pick from P1
    MoveL Offs(p1,0,0,100), v100,fine,tool;
    MoveL p1, v50, fine tool;
    Vacuum_ON;
    WaitUntil vacuum_ready=1\Timeout = 2;
    MoveL MoveL Offs(p1,0,0,100), v100,z10,tool;
    IF timeout=TRUE THEN
      Vacuum_ON;
      SetDO s1, 1;
    ELSE
      SetDO s1, 0;
    ENDIF
  CASE 5: ──────────────────────────────►  Place at P1
    MoveL Offs(p1,0,0,100), v100,fine,tool;
    MoveL p1, v50, fine tool;
    Vacuum_OFF;
    WaitUntil vacuum_ready=0\Timeout = 2;
    MoveL MoveL Offs(p1,0,0,100), v100,z10,tool;
```

```
    IF timeout=TRUE THEN
      SetDO s1, 1;
    ELSE
      SetDO s1, 0;
    ENDIF
    CASE 9:  ─────────────────────────────►  Go home
    MoveJ home, v100,z10,tool;
    CASE 10:  ────────────────────────────►  Start Vacuum
      SetDO doVacuum,1;
      WaitUntil vacuum_ready=1\Timeout = 2;
      IF timeout=TRUE THEN
        SetDO s1, 1;
      ELSE
        SetDO s1, 0;
      ENDIF
    CASE 11:  ────────────────────────────►  Stop Vacuum
      SetDO doVacuum,0;
      WaitUntil vacuum_ready=0\Timeout = 2;
      IF timeout=TRUE THEN
        SetDO s1, 1;
      ELSE
        SetDO s1, 0;
      ENDIF
ENDPROC
```

Figure 3.4 Application running on the robot controller (RAPID)

The application presented in Figure 3.4 uses the following variables:

- *command_ready* – this is a digital input signal used to specify that a valid command is ready to be read. This variable is defined as a USER IO SIGNAL in the robot system parameters
- *command* – group of four digital signals (d0, d1, d2 and d3) used to specify the command that should be executed. This variable is defined as a GROUP OF IO SIGNALS in the robot system parameters
- *status* – group of six digital output signals (s0, s1, s2, s3, s4 and s5) used to specify the robot status. This variable is also defined as a GROUP OF IO SIGNALS in the robot system parameters: s0 specifies if the robot is ready (1) or busy (0), s1 specifies if a command was correctly executed (0) or if there was any execution error (1), s2 is associated with the system ERROR OUTPUT ACTION, s3 is associated with the system AUTO OUTPUT action, s4 is associated with the system MOTOR ON OUTPUT action and s5 is associated with the system PROGRAM RUN OUTPUT action. Signals s2 to s5 are defined as SYSTEM OUTPUTS in the robot system parameters
- There are also four extra robot digital IO inputs, associated with the command of system actions *MOTOR ON, MOTOR OFF, PROGRAM*

RUN, and *PROGRAM STOP*. These signals were named *motor_on*, *motor_off*, *program_run* and *program_stop*, respectively, and are defined as SYSTEM INPUTS in the robot system parameters.

3.2.1.2 PLC Software

The PLC software was designed to operate as a server. Furthermore, the application is basically composed of a serial port interrupt and service routine that handles the communication with the PC, placing the received string on known memory locations. In this example, the received string is copied to the memory zone that starts with byte 90. Therefore, the following happens when a message is received:

VB90 – contains the number of bytes received
VB91 – contains the numeric code associated with that command
VB92 – contains parameter 1
…
VB92+N – contains parameter N
Note: In this example, the number of possible parameters is limited to 5, i.e., N = 5.

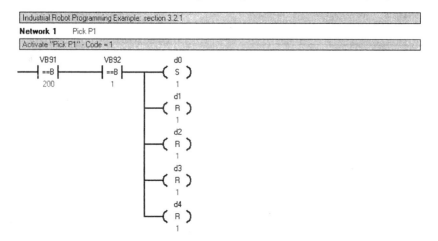

Symbol	Address	Comment
d0	Q0.0	Digital Output Q0.0 (24 Volts) -> to robot
d1	Q0.1	Digital Output Q0.1 (24 Volts) -> to robot
d2	Q0.2	Digital Output Q0.2 (24 Volts) -> to robot
d3	Q0.3	Digital Output Q0.3 (24 Volts) -> to robot
d4	Q0.4	Digital Output Q0.3 (24 Volts) -> to robot

a)

```
Industrial Robot Programming Example: section 3.2.1
Network 1     Pick P1
Activate "Pick P1" - Code = 1
LDB=    VB91,  200
AB=     VB92,  1
S       d0,  1
R       d1,  1
R       d2,  1
R       d3,  1
R       d4,  1
b)
```

Figure 3.5 Equation to activate action *"Pick P1"* using the SIEMENS programming suite for the *S7-200 PLC* model (*Step 7 Micro/Win 32 V4*) [3]: a – Ladder view, b – STL view

Furthermore, any PLC action will be triggered by a byte comparison between VB91 (byte carrying the received command numeric code) and the particular numeric code associated with that action, discriminating also the parameters associated with the command. For example, to activate the command *"Pick P1"* the following command string must be sent to the PLC:

$$200 \ 1 \ 0 \ 0 \ 0 \ 0$$

which results in making VB91 = 200 and VB92 = 1.

Consequently, the equation necessary to activate the action *"Pick P1"* is represented in Figure 3.5.

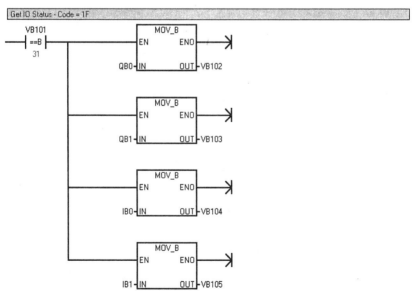

Figure 3.6 Ladder view of the "Get Robot Status" action on the PLC. Bytes VB100 to VB105 constitute an intermediate buffer used by the serial port service routine. Bytes QB0 and QB1 carry the state of all the digital outputs, and bytes IB0 and IB1 carry the state of all the digital inputs.

All the remaining actions are implemented in a similar way. Nevertheless, there is one special action that should return the robot status. This feature is obtained just by packing the actual status of all IO signals and sending it through the serial communication port, as the answer to the monitoring command *"Get Robot Status"* (code 1F) – Figure 3.6.

3.2.1.3 PC Software
The software developed to run on the PC provides the user interface to this setup. It is used to send the user selected commands to the PLC and to receive and present to the user the *"status"* information (Figure 3.7).

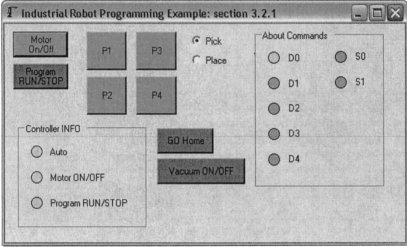

Figure 3.7 PC user interface

This simple application was coded using *Visual Basic .NET2005*. In the following (Figure 3.8) some aspects of the code associated with a few software buttons (actions) are revealed.

Motor On/Off	If robot_auto = 1 Then
	    com.Output = Chr(206) + Chr(0) + Chr(0) + Chr(0) + Chr(0) + Chr(0)
	Else
	    com.Output = Chr(207) + Chr(0) + Chr(0) + Chr(0) + Chr(0) + Chr(0)
	End If

Program RUN/STOP	If program_run = 1 Then
	    com.Output = Chr(208) + Chr(0) + Chr(0) + Chr(0) + Chr(0) + Chr(0)
	Else
	    com.Output = Chr(209) + Chr(0) + Chr(0) + Chr(0) + Chr(0) + Chr(0)
	End If

P1	If pick.Checked = True Then
	  com.Output = Chr(200) + Chr(1) + Chr(0) + Chr(0) + Chr(0) + Chr(0)
	End If
	If place.Checked = True Then
	  com.Output = Chr(201) + Chr(1) + Chr(0) + Chr(0) + Chr(0) + Chr(0)
	End If

Figure 3.8 Some actions available from the PC software

The actions "*Motor ON/OFF*" and "*Program RUN/STOP*" are obtained just by introducing a properly temporized *IO PULSE* on the relevant robot system input, which triggers those actions. Consequently, the PLC equation for the above mentioned actions is a simple *IO PULSE* obtained using the *PULSE* function or a TIMMER function. Figure 3.9 shows the *ladder* view for the "*Motor ON*" action and the corresponding timing.

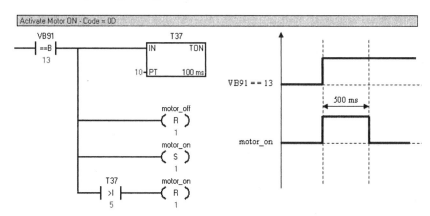

Symbol	Address	Comment
motor_off	Q0.6	Digital Output Q0.6 (24 Volts) -> to robot
motor_on	Q0.5	Digital Output Q0.5 (24 Volts) -> to robot

Figure 3.9 – *Ladder* view of the "*Motor ON*" action on the PLC, including a sketch of the timing of the obtained PULSE

To briefly summarize this section so far, a simple example was presented where a robot is used to pick-and-place objects from four pre-defined positions. An industrial PLC was used to interface the commanding PC with the robot controller. This example demonstrates the utilization of IO digital signals to design a simple communication and data interface for commanding and monitoring applications in industrial environments.

3.2.2 Using Fieldbuses

A fieldbus [4] is an industrial network used for distributed control, i.e., to use with systems in which the control function is distributed among the several components of the system. In fact, actual industrial components like sensors, actuators, drive systems, programmable controllers, etc., are equipped with powerful computing systems that enable the system designer to transfer part of the control software, associated with acquisition, control, and actuation tasks, to those systems, distributing in this way the overall control function. Consequently, the available fieldbuses were developed to provide a reliable platform to transmit IO data (digital and analog) between industrial PLCs and peripheral equipment, like sensors and actuators but also to established a low-level network with other PLCs and microprocessor-based programmable devices. Consequently, fieldbuses are mainly seen by users as a way to have remote IOs, i.e., a way to access remote sensors and actuators using a two-wire network, avoiding in this way a huge amount of cables and analog transmissions on the field (process) level. Furthermore, fieldbuses are also a reliable and convenient way to make application-oriented, low-level networks. There are several technical specifications available in the market, maintained by international and generally non-profit organizations, supported by the big majority of hardware manufacturers. Three of the most popular specifications will be covered here: *ProfiBus*, *CAN* and *DeviceNet* [4].

3.2.2.1 Profibus (Process FieldBus)
Profibus is probably the most popular type of fieldbus with more than 15 million installed devices as of 2006. It was developed in 1989 as a deliverable of a German research project, whose consortium was composed by several companies and research institutions.

Based on the real-time capable token-bus principle, *Profibus* handles multi-master and master-slave communications, allowing transfer rates up to 500 Kbits/s. *Profibus* is based on standards (the application, data, and physical layers are all standard) and enables reliable communication that distinguishes between confirmed and unconfirmed services allowing process communication, broadcast and real-time. Since *Profibus* is a master-slave pooling network with the ability to upload/download configuration data, it allows process synchronization of multiple devices on the network.

3.2.2.2 CAN (Controller Area Network)
CAN is a fast serial bus that was designed to provide an efficient, reliable, and very economical link between sensors and actuators. CAN uses a twisted pair cable to communicate at speeds up to 1Mbit/s with up to 40 devices. Originally developed to simplify the wiring in automobiles, its use has spread to machine and factory automation products. For example, SDS (*Smart Distribution System*) was developed by *Bosch* for networking most of the distributed electrical devices

throughout an automobile, initially for eliminating the large and expensive wiring harnesses at *Mercedes* (car manufacturer from Germany).

CAN provides standardized communication objects for process data, service data, network management, synchronization, time-stamping, and emergency messages. It is the basis of several sensor buses, such as *DeviceNet* (*Allen-Bradley*), SDS (*Smart Distribution System*) from *Honeywell* or CAL (*Can Application Layer*) from "*CAN in Automation Group*" (a group of about 300 international users and manufacturers). *CANOpen* is a family of profiles based on *CAN* which was developed within the "*CAN in Automation Group*". The extensive error detection and correction features of *CAN* may easily withstand the harsh physical and electrical environment presented by a car.

3.2.2.3 DeviceNet
DeviceNet is an extension of *CAN* adapted for critical factory networking purposes. At the next level are the "*control*" networks, which include *ControlNet*, developed by *Allen-Bradley* and also utilized by *Honeywell*, overlapping with some of the functionality provided by *Profibus-FMS* (*FieldBus Message Specification*). *Profibus-FMS* uses the same physical layer as *Profibus DP* (*Decentralized Peripheral*) but allows multi-master, asynchronous, peer-to-peer communication. *FMS* and *DP* can operate simultaneously on the same network. *ControlNet* was conceived as the ultimate high-level fieldbus network and was designed to meet several high performance automation and process control criteria. Of primary importance is the ability to communicate with each other being 100% deterministic, while achieving faster response than traditional master/slave poll/strobe networks.

Furthermore, *DeviceNet* is a simple, open networking solution that reduces the cost and time required to wire and install industrial automation devices, while providing interchangeability of components from multiple vendors. *DeviceNet* is a cost-effective solution for low-level industrial device networking and an effective way to provide access to the intelligence present in those devices. A *DeviceNet* network lets the user/programmer connect devices directly to shop floor controllers without hard-wiring each device into an I/O module. It is also used to:

- Reduce wiring and installation cost
- Reduce start-up time
- Significantly reduce downtime and the total cost of ownership with the aid of diagnostics, Auto Device Replacement, and other time- and cost-saving features
- Support standard and safety applications on the same wire
- Benefit from an open network
- Control, configure, and collect data on a single network

Consequently, using a fieldbus is not significantly different if compared to regular IO, since the same logic of encoding commands and parameters is used, utilizing the IO signals/bits like a data bus. Nevertheless, fieldbuses use high bit rates over a reduced number of wires (normally a twisted-pair cable), which is an enormous advantage for industrial utilization since it allows a considerable reduction in the number of wires within the system. Other than that, since a fieldbus can accommodate a big number of remote IOs, it is easier to implement a messaging protocol to handle the necessary commands and related parameters, events, and monitoring tasks. In fact, many of the fieldbus consortiums developed their own protocols and consequently the user can choose between his own protocol, or the one available from the specific technology adopted.

Currently there is a debate about using *Ethernet* with predictable timing (deterministic and robust) for "*fieldbus type*" operations, i.e., penetrating deep into the factory network hierarchy, down to the I/O level. This is justified by the fact that Ethernet is a network commonly available on the shop floor and used for many operations between controllers and computers. A decade ago, no serious design engineer would have suggested using *Ethernet* for networking shop floor devices.

Ethernet, the technology for office automation, was developed more than three decades ago as a high-speed serial data-transfer network. It has become a worldwide standard and is now the most widely used *Local Area Network* (LAN). More than 85% of all installed network connections in the world are *Ethernet*. But it was deliberately ignored for industrial applications, and for good reasons: Its lack of determinism and robustness made it feeble and not suitable for the shop floor. Nevertheless, with time and research things changed, and today the scene is considerably different. In fact, over the past few years there have been many enhancements to the Ethernet standard, especially in areas of determinism, speed, and message prioritization. So there is no longer any reason why *Ethernet* cannot be used to build deterministic fieldbus networks that are cost-effective and open. And since *Ethernet* is already the network choice for business computing, its presence at the control level will facilitate the integration of low-level data with high-level applications.

Another good reason why manufacturers are looking at *Ethernet* is the coming explosion of shop floor data traffic. As smart sensors and various devices on the shop floor consume the available bandwidth over the next few years, manufacturing plant information generated by PLCs and control systems is expected to increase from 10 to 30 times the current level. *Ethernet*, with its Internet-friendly TCP/IP protocol, is ideally positioned. It is popular, sinking in price and being propelled by utter market demand.

Nevertheless, this scenario makes some of the PLC manufacturers uncomfortable. Even the recently arrived fieldbus systems are beginning to feel threatened by *Ethernet*. Furthermore, the *DeviceNet*, *Profibus* and *Foundation Fieldbus* protocols are all available or in development as application layers for *Ethernet*. And most PLCs now offer *Ethernet* as a standard networking option in addition to their fieldbus of choice. *High Speed Ethernet* (HSE) is a 100 Mbit/s *Ethernet* standard that uses the same protocol and objects as *Foundation Fieldbus H1*, on TCP/IP.

The new generation of *Ethernet* is called *Gigabit Ethernet*, which is capable of 1 Gbits/sec. This will bridge the gap between the necessity of industrially hardened wiring capability and the growing need for process data via business LANs and the Internet. Most companies cannot afford to have a *DeviceNet* or *Profibus* specialist on their technical staff. Even if a company could afford such a person, it is unlikely that fieldbus would be their specialty. However, almost every company has a network administrator who is well versed and specialized in the *Ethernet* protocol, making *Ethernet* even more attractive for industrial control.

In this book, *Ethernet* and TCP/IP network protocols are used extensively for several types of tasks:

1. To command distributed systems from remote computers
2. To supervise and monitor operation of the manufacturing systems
3. To exchange data, configuration setup, etc., with peripheral devices (sensors and actuators, for example)
4. To monitor and supervise operation of the remote systems, including controllers, sensors, actuator modules, etc
5. To program peripheral devices (sensors and actuators) and/or adjust their behavior
6. To receive events (asynchronous calls) from peripheral devices with data, warnings, or errors

3.3 Data Protocols and Connections

The challenges posed by any robotic manufacturing system are similar and independent of the particular application under study. Consequently, the software architecture [5-7] presented in this book was designed to be used with generic robotic manufacturing cells that may include several types of equipment like robot manipulators, mobile robots, PLCs (programmable controllers), CNC machines, vision systems and several types of sensors, *etc.* Usually these systems use different programming languages, even when the manufacturer is the same. It is then very difficult to make adjustments to the cell functionality, or adapt it to new requirements posed by the introduction of a new product or by changes introduced in existing products. Several research and technical efforts have been carried out to overcome these problems. Many of those efforts point to solutions that consider the development of general programming languages that could be used with any equipment, relying on individual interpreters to generate the specific code for any equipment.

Nevertheless, recent research works show that it is desirable to have a flexible environment and still program each machine using its own language. The reason is simple: a general syntax means introducing generalizations and simplifications that tend to limit the potentiality of the equipment. Consequently, some parameterization is not used, special non-grouped functions are not used, and the generated code takes always a uniform structure which may not be the best for all machines.

The idea presented here is rather different, being an alternative to the solutions presented in the literature, and also for the software products truly distributed available on the market. The basic idea is to define for each individual machine a collection of software functions that expose all its basic operational features. That objective requires local processing capabilities, availability of communication channels, and support for the standard technologies used when implementing the services installed on the individual machines. Since the vast majority of the current robotics and automation (R&A) equipment meets these requirements fully, this is not a serious limitation. Also, the above-mentioned services are to be offered through a local network, on a distributed software framework based on the client-server model. Furthermore, using those services from the remote client computer to build controlling and inspection applications can be performed from any platform (*UNIX*, *Linux*, *Win32-DC*OM, *etc.*), using standard programming languages (*C*, *C++*, *C#*, *Visual Basic, etc.*).

Several approaches can be used and are currently available from various robot manufacturers, with specific details and implementations. Nevertheless, the following objectives are pursued by any of the above-mentioned software architectures:

1. Be able to represent the robot manipulator's motion based on the kinematic and dynamic models, but also based on real-time data coming from the real robot. That can be done using available mathematical and graphical software packages, like *Matlab* for example. This latest objective clearly indicates the need to access robot motion and status information in real-time from the mathematical package

2. Be able to develop applications to explore remotely the entire installation (robot and welding application, for example) using standard programming languages (*C*, *C++*, *C#*, *Visual Basic, etc.*)

3. Be able to integrate and explore intelligent sensors used to obtain information from the process under control

4. Enable users to explore the advanced programming capabilities of actual robot controllers, namely the local programming capabilities, based on a dedicated programming language complemented by extensive libraries of functions, and the optimized manipulation capabilities based on trajectory planning software that takes advantages of optimized kinematic and dynamic models

5. Enable users to build flexible manufacturing cells, which leads to the ability to explore the available industrial data network, and to distribute software to the various components of the system, as well as the capacity to build remote software applications to control and monitor industrial manufacturing cells

6. Develop advanced *Human Machine Interface* (HMI) solutions to operate with industrial systems, hiding from the users all the tricky details about

implementation, allowing them to focus on the operational details, *i.e.*, to focus on how systems work and how they can be explored efficiently

7.   Provide ways that could allow developers to focus on the important things about the application they are building: the control algorithm, program functionality, and HMI. All the details related to communications, sensor integration, *etc.*, should be hidden from the user

Taking into consideration these objectives, the following programming models are required:

1.   **Client-server model:** There should be server code running on each cell equipment, namely on the robot controllers and coordinating PLCs, that could receive calls from the remote client computers, execute the commands and return the results

2.   **Remote procedure calls**: This is the most usual method used to implement communications between a client and a server on a distributed environment. The client makes a call to a non-local function and the selected RPC mechanism configures the call so that the proper computer, server program and function are addressed, adding the necessary network headers. The server program, running on the server machine, receives the call, executes the selected function, and returns the results obtained to the client computer

3.   **IPC socket connections**: Another approach is to use TCP or UDP sockets to make the interprocess (IPC) and intersystem communication, defining a messaging mechanism to send commands and obtain process data

4.   **Data sharing**: Most of the services require data sharing, files and databases between the client and the server. Consequently, the mechanism provided by the RPC technology to implement data sharing must be used

Another important thing to consider is the need to interface intelligent sensors with the system. The most easy and portable way to do that is to build software components that implement the methods, properties and data structures necessary to configure and use the sensor. Consequently, a technology to implement software components is also needed. The basic architecture presented in Figure 3.10 details all these requirements.

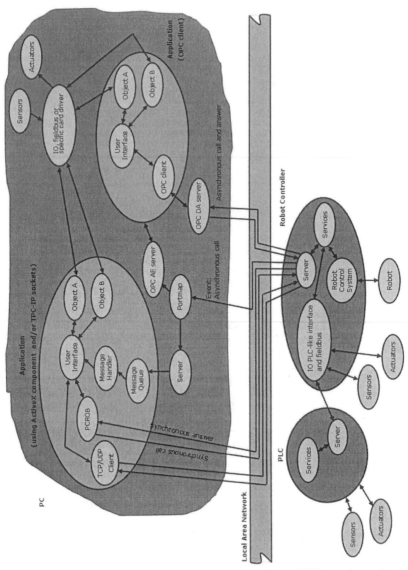

Figure 3.10 Software architecture used (depicting several possibilities: using software components, using RPC sockets, using TCP/IP sockets and OPC – *OLE for Process Control*)

Sockets provide point-to-point, two-way communication between two processes. Sockets are very versatile and are a basic component of interprocess and intersystem communication. A socket is an end point of communication to which a name can be bound. It has a type and one or more associated processes.

Sockets exist in communication domains (*families*). A socket domain is an abstraction that provides an addressing structure and a set of protocols. Sockets connect only with sockets in the same domain. Several domains are identified and can be used to communicate between processes on a single system, like other forms of IPC.

Sockets can also be used to communicate between processes on different systems. The socket address space between connected systems is called the Internet domain, and in that case the communication uses the TCP/IP Internet protocol suite.

Socket types define the communication properties visible to the application. Processes communicate only between sockets of the same type. There are several types of socket:

Stream socket - provides two-way, sequenced, reliable, and unduplicated flow of data with no record boundaries. Stream sockets operate much like a telephone conversation. The socket type is *SOCK_STREAM*, which, in the Internet domain, uses *Transmission Control Protocol* (TCP).

Datagram socket - supports a two-way flow of messages, not necessarily sequenced (messages can appear in a different order), and unreliable flow of data with record boundaries. Datagram sockets operate much like passing letters back and forth in the mail. The socket type is *SOCK_DGRAM*, which, in the Internet domain, *uses User Datagram Protocol* (UDP).

Sequential packet socket - provides a two-way, sequenced, reliable, connection, for datagrams of a fixed maximum length. The socket type is *SOCK_SEQPACKET*. No protocol for this type has been implemented for any protocol family.

Raw socket - provides access to the underlying communication protocols. These sockets are usually datagram-oriented, but their exact characteristics depend on the interface provided by the protocol.

In this book, we use *stream sockets* (for TCP client-server connections) and *datagram sockets* (for UDP client-server connections. Figure 3.11 shows the code used to open a socket on a TCP client application.

```
Private Shared Function C_Sock(ByVal server As String, ByVal port As Integer) As Socket
   Dim s As Socket = Nothing
   Dim hostEntry As System.Net.IPHostEntry = Nothing
   Dim address As IPAddress
   address = IPAddress.Parse(server)
   Dim endPoint As New IPEndPoint(address, Integer.Parse(port))
   Dim tempSocket As New
   Socket(AddressFamily.InterNetwork, SocketType.Stream, ProtocolType.Tcp)
   Try
     tempSocket.Connect(endPoint)
     If tempSocket.Connected Then
        s = tempSocket
     End If
   Catch e As Exception
     Return s
   End Try
   Return s
End Function
```

Figure 3.11 Code used to open a TCP socket connection (using *Visual Basic .NET 2005*)

Admitting that there's a TCP socket server running on the robot controller, as an independent task (process), which receives remote commands through the open socket, executes them, and returns the correspondent results, Figure 3.12 shows what a simple "*motor_on*" command should look like.

```
server_name = ip.Text
server_port = port.Text
s = ConnectSocket(server_name, server_port)
If s Is Nothing Then
   ans_robot.Text() = "Error connecting to robot, master."
Else
   Dim bytesSent As [Byte]() = Nothing
   bytesSent = ascii.GetBytes("motor_on")
   s.Send(bytesSent, bytesSent.Length, 0)
   bytes = s.Receive(bytesReceived, bytesReceived.Length, 0)
   ans_robot.Text() = Encoding.ASCII.GetString(bytesReceived, 0, bytes)
   s.Close()
   If Encoding.ASCII.GetString(bytesReceived, 0, bytes) = "0" Then
      ans_robot.Text() = "Motor on, master."
      cstate.Text() = "Motors ON"
   Else
      ans_robot.Text() = "Error executing, master."
   End If
End If
```

Figure 3.12 Sample code used to command the action "*motor_on*" with TCP sockets (using *Visual Basic .NET 2005*)

This code will be used later in this book with several examples that explore the utilization of *stream* and *datagram* sockets to command industrial robotic applications.

3.3.1 RPC – Remote Procedure Calls

A *remote procedure call* (RPC) is a facility that a software application can use to request a service from a program located in another computer of the network without having to understand network details. (A *procedure call* is also sometimes known as a *function call* or a *subroutine call*.) RPC uses the well known client-server model. The requesting program is the *client* and the service-providing program is the *server*. Like a regular or *local procedure call*, an RPC is a synchronous operation requiring the requesting program to be suspended until the results of the remote procedure are returned. However, the use of *lightweight processes,* or threads that share the same address space, allows multiple RPCs to be performed concurrently.

When the software statements that use RPCs are compiled into an executable program, a *stub* is included in the compiled code that acts as the representative of the remote procedure code. When the software is executed and the procedure call is issued, the *stub* receives the request and forwards it to a client runtime program in the local computer. The client runtime program knows how to address the remote computer and server application, and sends the message across the network that requests the remote procedure. Similarly, the server includes a runtime program and *stub* that interface with the remote procedure itself. Results are returned the same way.

There are several RPC models and implementations. A popular model and implementation is the *Open Software Foundation's Distributed Computing Environment* (DCE). *The Institute of Electrical and Electronics Engineers* (IEEE) defines RPC in its *ISO Remote Procedure Call Specification*, ISO/IEC CD 11578 N6561, ISO/IEC, November 1991.

RPC is a powerful technique for constructing distributed, client-server based applications. It is based on extending the notion of conventional or local procedure calling, so that the called procedure need not exist in the same address space as the calling procedure. The two processes may be on the same system, or they may be on different systems with a network connecting them. By using RPC, programmers of distributed applications avoid the details of the interface with the network. The transport independence of RPC isolates the application from the physical and logical elements of the data communications mechanism and allows the application to use a variety of transports.

RPC makes the client/server model of computing more powerful and easier to program. When combined with the *ONC RPCGEN* protocol compiler, clients transparently make remote calls through a local procedure interface.

Consequently, the robot controller software works as a server, exposing to the client a collection of RPC services that constitute its basic functionality. Those services, offered by the RPC servers running on the robot controller, include the variable access services, files and programs management services, and robot status and controller-state management and information services. To access those services, the remote computer (*client*) issues properly parameterized remote procedure calls to the robot controller (*server*) through the network.

Considering, for example, the S4CPLUS robot controller from ABB Robotics, it's possible to extend the RPC services available in the robot controller adding user functionality to the system. The ABB implementation is based on a messaging protocol developed by ABB called RAP (*remote application protocol*) [8], which is an *application specific protocol* (ASP) of the OSI application level. The messaging protocol RAP defines the necessary data structures and message syntax of the RPC calls used to explore the RPC services available in the controller.

These services were implemented using the standard and open source RPC specification SUN RPC 4.0, a collection of tools developed by the *SUN Microsystems Open Network Group* (ONC) [2]. Consequently, to implement the client calls, the *ONC SUN RPC 4.0* specification and tools were also used. This package includes a compiler (*rpcgen*), a *portmaper* and a few useful tools like *rpcinfo*. The Microsoft RPC implementation uses another standard defined by *Digital Corporation* named *OSF/DCE*, which is not compatible with the SUN RPC standard. The package used to build the client software was a port to *Windows NT/2000/XP*, equivalent to the original version that was built to *UNIX* systems, although a few functions were slightly changed to better suit the needs without compromising compatibility with client and server programs developed with the *SUN RPC* package. The port was compiled using the *Microsoft Visual C++ .NET 2003* compiler.

From all the RPC services available in the robot controller, the ones really needed to implement the software architecture depicted in Figure 3.10 are the variable access services. Nevertheless, calls to the other services were implemented for completeness. The procedure is simple and based on the XDR (*extended data representation*) file obtained by defining the data structures, the service identification numbers, and the service syntax specified by the RAP protocol. That file is compiled by the *rpcgen* tool, generating the basic calls and data structure prototypes necessary to invoke the RPC services available from the robot controller. The necessary code was added to each basic function and packed into an *ActiveX* software component named *PCROBNET2003/5* [5-7]. The complete set of functions included in this object is listed in Table 3.3.

Although this software component was built using the DCOM/OLE/ActiveX object model, it does not run the *Microsoft RPC* implementation but instead the already mentioned *SUN RPC 4.0* port to the *Win32* API.

Table 3.3 Methods and properties of the software component PCROB NET2003/5

Function	Brief description
open	Opens a communication line with a robot (RPC client)
close	Closes a communication line
motor_on	Go to run state
motor_off	Go to standby state
prog_stop	Stop running program
prog_run	Start loaded program
prog_load	Load named program
prog_del	Delete loaded program
prog_set_mode	Set program mode
prog_get_mode	Read actual program mode
prog_prep	Prepare program to run (program counter to begin)
pgmstate	Get program controller state
ctlstate	Get controller state
oprstate	Get operational state
sysstate	Get system state
ctlvers	Get controller version
ctlid	Get controller ID
robpos	Get current robot position
read_xxxx	Read variable of type xxxx (there are calls for each type of variable defined in RAPID)
read_xdata	Read user-defined variables
write_xxx	Write variable of type xxxx (there are calls for each type of variable defined in RAPID)
write_xdata	Write user-defined variables
digin	Read digital input
digout	Set digital output
anain	Read analog input
anaout	Set analog output

To use a remote service, the computer running the client application needs to make properly parameterized calls to the server computer, and receive the execution result. Two types of services may be considered: synchronous and asynchronous. The synchronous services return the execution result as the answer to the call.

Consequently, the general prototype of this type of call is:

$$\textbf{short } status \; call\_service\_i \; (\textbf{struct } parameters\_i, \textbf{struct } answer\_i)$$

where *status* returns the service error codes (zero if the service returns without errors, and a negative number identifying the error otherwise), *parameters_i* is the data structure containing the service parameters and *answer_i* is the data structure that returns the service execution results.

The asynchronous services, when activated, return answers/results asynchronously, *i.e.*, the remote system should also make remote procedure calls to the client system when the requested information becomes available or when the specified event occurs (system and controller state changes, robot program execution state change, IO and variable events, *etc.*). Those calls may be named events or spontaneous messages, and are remote procedure calls issued to all client computers that made the correspondent subscription, *i.e.*, made a call to the subscription service properly parameterized specifying the information wanted. To receive those calls, any remote client must run RPC servers that implement a service to receive them (Figure 3.13). The option adopted was to have that server broadcast registered messages inside the operating system, enabling all open applications to receive and process that information by filtering its message queue.

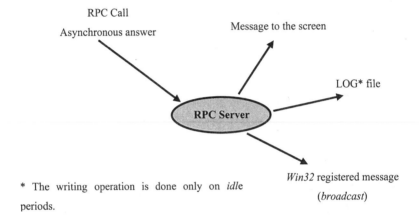

RPC Call

Asynchronous answer

Message to the screen

LOG* file

RPC Server

Win32 registered message

(*broadcast*)

* The writing operation is done only on *idle* periods.

Figure 3.13 Functionality of the RPC server necessary to receive spontaneous messages

As mentioned already, the variable access services allow access to all types of variables defined in the robot controller. Using this service, and developing the robot controller software in a convenient way, it is possible to add new services to the system. In fact, that possibility may be achieved by using a simple SWITCH-CASE-DO cycle driven by a variable controlled from the calling (client) remote computer:

```
switch (decision_1)
{
    case 0: call service_0; break;
    case 1: call service_1; break;
    case 2: call service_2; break;
    ...
    case n: call service_n; break;
}
```

This server runs on the robot controller, making the process of adding a new service a simple task. The programmer should build the procedure (routine) that implements the new functionality, and include the call to that procedure in the server cycle by identifying it with the specific number of the control variable.

This is not far from what is done with any RPC server; the *svc_run* function, used in those programs is a *SWITCH-CASE-DO* cycle that implements calls to the functions requested by the remote client. With this type of structure it is straightforward to build complex and customer functions that can be offered to the remote client. Furthermore, with this approach it's still possible to use the advanced capabilities offered by the robot controller, namely the advanced functions designed to control and setup the robot motion and operation. Examples exploring this facility are presented and discussed in this chapter (sections 3.4 to 3.6).

3.3.2 TCP/IP Sockets

One of the most interesting ways to establish a network connection between computer systems is by using TCP/IP sockets. This is a standard client-server procedure, not dependent on the operating system technology used on any of the computer systems, requiring only the definition of a proper messaging syntax to be reliable and safe. The user-defined messaging protocol should specify the commands and data structures adapted to the practical situation under study.

The TCP/IP protocol suite is based on a four-layer reference model. All protocols that belong to the TCP/IP protocol suite are located in the top three layers of this model.

As shown in Figure 3.14, each layer of the TCP/IP model corresponds to one or more layers of the seven-layer *Open Systems Interconnection* (OSI) reference model proposed by the *International Standards Organization* (ISO).

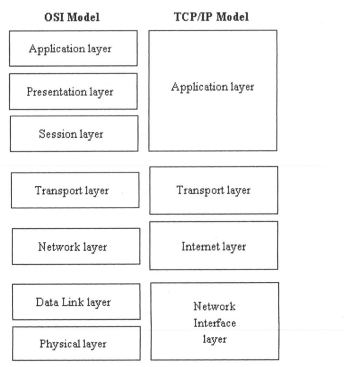

Figure 3.14 Correspondence between the OSI Model and the TCP/IP Model in terms of layers.

Table 3.4 Services performed at each layer of the TCP/IP Model

Layer	Description
Application	Defines the TCP/IP application protocols and how the host programs interface with transport layer services to use the network
Transport	Provides communication session management between host computers. Defines the level of service and the status of the connection used when transporting data
Internet	Packages data into IP datagrams, which contain source and destination address information that is used to forward the datagrams between hosts and across networks. Performs routing of IP datagrams
Network interface	Specifies details of how data is physically sent through the network, including how bits are electrically signaled by hardware devices that interface directly with a network medium, such as coaxial cable, optical fiber, or twisted-pair copper wire

The types of services performed at each layer within the TCP/IP model are described in more detail in Table 3.4.

Transmission control protocol (TCP) is a required TCP/IP standard defined in RFC 793, "*Transmission Control Protocol (TCP)*" that provides a reliable, connection-oriented packet delivery service. The *transmission control protocol*:

- Guarantees delivery of IP datagrams
- Performs segmentation and reassembly of large blocks of data sent by programs
- Ensures proper sequencing and ordered delivery of segmented data
- Performs checks on the integrity of transmitted data by using checksum calculations
- Sends positive messages depending on whether data was received successfully. By using selective acknowledgments, negative acknowledgments for data not received are also sent
- Offers a preferred method of transport for programs that must use reliable session-based data transmission, such as client/server database and e-mail programs

TCP is based on point-to-point communication between two network hosts. TCP receives data from programs and processes this data as a stream of bytes. Bytes are grouped into segments that TCP then numbers and sequences for delivery.

Before two TCP hosts can exchange data, they must first establish a session with each other. A TCP session is initialized through a process known as a *three-way handshake*. This process synchronizes sequence numbers and provides control information that is needed to establish a virtual connection between both hosts.

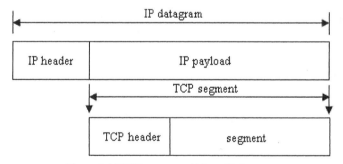

Figure 3.15 TCP segment within an IP datagram

Once the initial *three-way handshake* completes, segments are sent and acknowledged in a sequential manner between both the sending and receiving hosts. A similar handshake process is used by TCP before closing a connection to verify that both hosts are finished sending and receiving all data.

TCP segments are encapsulated and sent within IP datagrams, as shown in Figure 3.15

3.3.2.1 TCP Ports

TCP ports use a specific program port for delivery of data sent by using the *transmission controlpProtocol*. TCP ports are more complex and operate differently from UDP ports.

While a UDP port operates as a single message queue and the network endpoint for UDP-based communication, the final endpoint for all TCP communication is a unique connection. Each TCP connection is uniquely identified by dual endpoints. Each single TCP server port is capable of offering shared access to multiple connections because all TCP connections are uniquely identified by two pairs of IP address and TCP ports (one address/port pairing for each connected host).

The server side of each program that uses TCP ports listens for messages arriving on their well-known port number. All TCP server port numbers less than 1024 (and some higher numbers) are reserved and registered by *the Internet Assigned Numbers Authority* (IANA).

3.3.3 UDP Datagrams

The *User Datagram Protocol* (UDP) is a TCP/IP standard defined in RFC 768, "*User Datagram Protocol (UDP)*". UDP is used by some programs instead of TCP for fast, lightweight, unreliable transportation of data between TCP/IP hosts.

UDP provides a connectionless datagram service that offers best-effort delivery, which means that UDP does not guarantee delivery or verify sequencing for any datagrams. A source host that needs reliable communication must use either TCP or a program that provides its own sequencing and acknowledgment services.

UDP messages are encapsulated and sent within IP datagrams, as shown in 3.16.

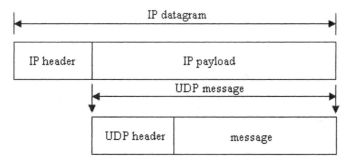

Figure 3.16 UDP message within an IP datagram

3.3.3.1 UDP Ports

UDP ports provide a location for sending and receiving UDP messages. A UDP port functions as a single message queue for receiving all datagrams intended for the program specified by each protocol port number. This means UDP-based programs can receive more than one message at a time.

The server side of each program that uses UDP listens for messages arriving on their well-known port number. All UDP server port numbers less than 1024 (and some higher numbers) are reserved and registered by the Internet *Assigned Numbers Authority* (IANA).

Each UDP server port is identified by a reserved or well-known port number.

3.4 Simple Example: Interfacing a CCD Camera

The example presented in this section demonstrates the utilization of TCP/IP sockets (stream type or TCP sockets) to command an industrial robot and to interface with a CCD camera (a common USB Webcam). The example will be presented in detail with the objective of allowing the reader to explore further from the concepts and ideas presented.

Basically the system is composed of the following components (Figure 3.17):

- Industrial robot manipulator ABB IRB140 equipped with the new IRC5 robot controller
- Pneumatic tool equipped with a vacuum cup
- Working table and several working pieces
- Webcam used to obtain the number of pieces present in the scene and their respective positions
- Pocket PC running the *Windows Mobile 2005* operating system

Figure 3.17 Setup for this example showing: Robot manipulator, Webcam, laptop running the Webcam TCP/IP server, and the commanding Pocket PC

The user is supposed to control the setup using the Pocket PC, namely:

- Change the robot state and start/stop program execution
- Interface with the Webcam, request the camera to identify the number of objects present in the scene and return their actual positions (Figure 3.18)
- Command the robot to pick-and-place the selected objects

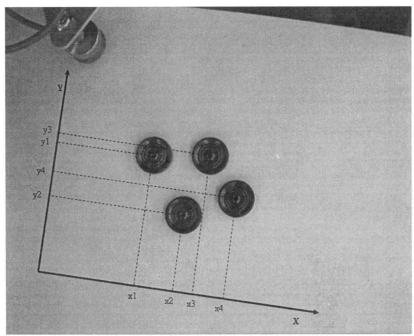

Figure 3.18 Returning the position of the objects present in the working scene based on the computed Cartesian position (x,y)

To build a solution to execute the above specified functions, it is necessary to handle several different subjects:

- Build a TCP/IP socket server to run on the robot controller. The server should implement a collection of services equivalent to the ones listed in Table 3.3
- Build an application to handle the webcam permitting to use it as a sensor and return the number of objects in the scene and their position. That application must run on a machine accessible from the network
- Build an application to command the setup offering a *human-machine interface* (HMI) facility

The following section provides a closer look at these three software packages.

3.4.1 Robot Controller Software

The robot controller runs two very different types of applications:

- The socket server used to implement the remote services and offer them to the remote clients
- The application that executes the commanded pick-and-place operations

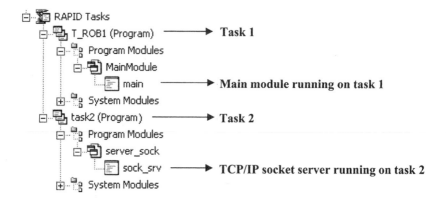

Figure 3.19 View of the tasks available on the system using *RobotStudio Online* (ABB)

The above mentioned applications are different applications in terms of objectives and requirements. Consequently, since the robot control system is a multitasking system, each of them was designed to run in their own task (process) – see Figure 3.19.

A TCP/IP socket server can work like a switch-case-do cycle driven by the received message. The first word of the received message, named "*command*", can be used to drive the cycle and discriminate the option to execute, implementing in this way the services it was designed to offer. Consequently, the TCP/IP server (*sock_srv*, running on task 2) should have a basic structure like the one represented in Figure 3.20.

```
PROC sock_srv()
  SocketCreate server_socket;
  SocketBind server_socket, "172.16.0.89", 2004;
  SocketListen server_socket;
  WHILE TRUE DO
    SocketAccept server_socket, client_socket;
    SocketReceive client_socket \Str := receive_string;
    extract_INFO_from_message (command, parameter{i});
    TEST command
      Case "motor_on"
        motor_on(result);
        SocketSend client_socket, result;
      Case "motor_off"
        motor_off(result);
        SocketSend client_socket, result;
      Case "write_num"
```

```
      write_num(parameter1, parameter2, result);
      SocketSend client_socket, result;
    Case "read_num"
      read_num(parameter1, result);
      SocketSend client_socket, result;
...
   ENDTEST
   SocketClose client_socket;
   ENDWHILE
   ERROR_HANDLER;
ENDPROC
```

Figure 3.20 Basic structure of the TCP/IP socket server running on the robot controller

The server briefly presented in Figure 3.20 implements basically the same functionality listed in Table 3.3. Furthermore, the command strings have a simple structure:

command parameter_1 parameter_2 ... parameter_N

i.e., the command string starts with a word representing the "*command*" (used by the server to discriminate what is the service the user wants to execute), followed by other words corresponding to the "*parameters*" associated with the "*command*". For example:

Action	Command String
Motor_ON	"*motor_on*"
Motor_OFF	"*motor_off*"
Read_num	"*read_num variable_name*"
Write_num	"*write_num variable_name value*"
Program_start	"*program_start module*"
Program_stop	"*program_stop module*"
...	

where "*variable_name*" is the name of the variable to read, "*value*" is the new value to assign to the variable, and "*module*" is the name of the module to start or stop.

3.4.2 Webcam Software

The application designed to handle the Webcam (Figure 3.17) also works as a TCP/IP server. The reason is simple, the Webcam works here as a sensor used to obtain two types of information: the number of objects and their respective position. Consequently, it is important to be able to address the sensor as an independent entity on the network, and simply command it to return the required information. One simple way to do that is to also adopt a client-server model for the Webcam software, using TCP/IP sockets to implement it. The software development package used here to add image processing capabilities to the developed software was *LabView* from *National Instruments*. Consequently, the complete application was built on *Labview*, including the TCP/IP socket implementation (Figure 3.21).

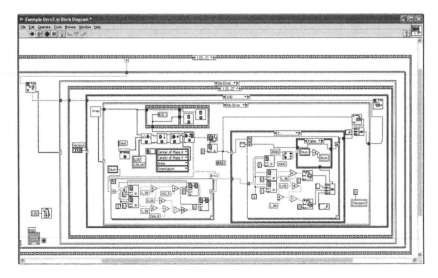

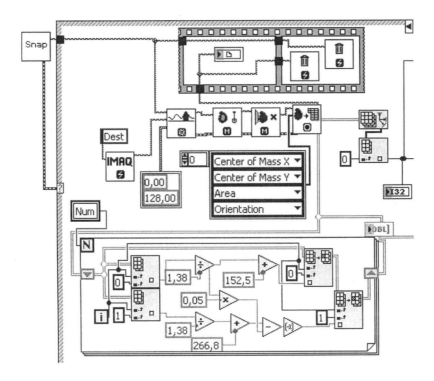

Figure 3.21 *Labview* Vi of the Webcam software (using IMAQ for *LabView*): a – complete VI; b – detail of part of the VI (feature computation)

The Webcam used here is a simple commercial USB Webcam (Figure 3.22) which must be installed on the machine running the above *Labview* mentioned Webcam application.

Figure 3.22 Webcam used in this application (*i-C@AM from Liftech Inc.*)

The TCP/IP server handling the Webcam software listens for commands on a specified IP address and port number. When a connection is accepted, the server responds to the following command:

Command - "*camera get objects*"

After receiving the command correctly the server acquires a frame from the Webcam and runs the image processing routine developed for this application. The routine identifies the objects in the captured frame, and for each object computes the center of mass. The TCP/IP client receives the following information:

- Number of objects identified
- Center of mass of each of the identified objects

The answer is sent through the open socket on a string with the following syntax:

$$number\_\#x1\_y1\#x2\_y2\#...\#xN\_yN\#$$

where "*number*" is the number of objects identified and *(xi, yi)* is the position of each of the objects. For example, for the scene presented in Figure 3.18:

command from client: "*camera get objects*"
answer from server: *4_#x1_y1#x2_y2#x3_y3#x4_y4#*

3.4.3 Remote Client

The objective of this application is to implement the human-machine interface with the user, providing the resources to enable the user/programmer to command the robot to pick-and-place the existing objects identified by the software associated with the Webcam. Basically, the application can run on any machine with access to the network. For this particular application, a Pocket PC (PPC) running *Windows Mobile 2005* was chosen since the PPC platform is powerful and very interesting for portable HMI applications, namely when a wireless network is available (Figure 3.23).

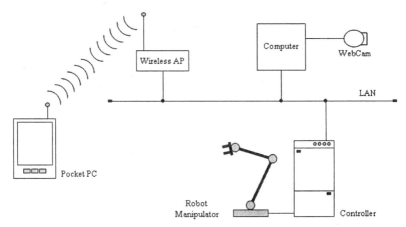

Figure 3.23 Overview of the setup used in this application

In the following material, the code of the client application will be briefly presented, showing in detail a few selected and representative functions. Figure 3.24 shows the screen of the developed PPC application used to connect to the TCP/IP server running on the robot controller and change the robot operating state.

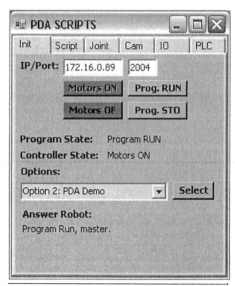

Figure 3.24 PPC screen to initialize robot operation and select program option

This is the code associated with the action "Motors ON" (Figure 3.24):

```
server_name = ip.Text
server_port = port.Text
sock = ConnectSocket(server_name, server_port)
If sock Is Nothing Then
   ans_robot.Text() = "Error connecting to robot, master."
Else
   Dim bytesSent As [Byte]() = Nothing
   bytesSent = ascii.GetBytes("motor_on")
   sock.Send(bytesSent, bytesSent.Length, 0)
   bytes = sock.Receive(bytesReceived, bytesReceived.Length, 0)
   ans_robot.Text() = Encoding.ASCII.GetString(bytesReceived, 0, bytes)
   moff.Enabled = True
   mon.Enabled = False
   prun.Enabled = True
   pstop.Enabled = True
   sel.Enabled = True
   sock.Close()
   If Encoding.ASCII.GetString(bytesReceived, 0, bytes) = "0" Then
      ans_robot.Text() = "Motor on, master."
      cstate.Text() = "Motors ON"
   Else
      ans_robot.Text() = "Error executing, master."
   End If
End If
```

The code presented above simply opens the socket, sends the commanding string, and processes the answer. This code is associated with the software button "*Motor ON*" in Figure 3.24.

To give another example, the code associated with the action "*Program RUN*" (Figure 3.24) is presented below:

```
Server_name = ip.Text
server_port = port.Text
sock = ConnectSocket(server_name, server_port)
If s Is Nothing Then
   ans_robot.Text() = "Error connecting, master."
Else
   Dim bytesSent As [Byte]() = Nothing
   bytesSent = ascii.GetBytes("program_start_main")
   sock.Send(bytesSent, bytesSent.Length, 0)
   bytes = sock.Receive(bytesReceived, bytesReceived.Length, 0)
   ans_robot.Text() = Encoding.ASCII.GetString(bytesReceived, 0, bytes)
   sock.Close()
   If Encoding.ASCII.GetString(bytesReceived, 0, bytes) = "0" Then
      ans_robot.Text() = "Program Run, master."
      pstate.Text() = "Program RUN"
   Else
      ans_robot.Text() = "Error executing, master."
   End If
End If
```

The interface with the Webcam is done through the screen window represented in Figure 3.25. Using this window, the user can command the camera to return the information about the objects in the scene. All the returned positions are listed in the list-box present in the interface (Figure 3.25) for the user to select the one he wants to use for the pick-and-place operation.

The code below details the implementation of the action "*Get Webcam Picture*" (Figure 3.25):

```
Dim msg_received As String
Dim indx As Integer
Dim num_obj As Integer
Dim index As Integer
sock = ConnectSocket(ip2.Text.ToString, port2.Text.ToString)
If sock Is Nothing Then
   ans_robot_3.Text() = "Error connecting to CCD, master."
Else
   Dim bytesSent As [Byte]() = Nothing
   bytesSent = ascii.GetBytes("camera get objects")
```

```
If s.Available <> 0 Then
    bytes = sock.Receive(bytesReceived, bytesReceived.Length, 0)
    MsgBox("ok, buffer cleared.")
End If
sock.Send(bytesSent, bytesSent.Length, 0)
bytes = sock.Receive(bytesReceived, bytesReceived.Length, 0)
list_cam.Items.Clear()
msg_received = Encoding.ASCII.GetString(bytesReceived, 0, bytes)
If msg_received <> "0_#no objects" Then
    indx = msg_received.IndexOf("#")
    num_obj = msg_received.Substring(0, indx - 1)
    n_obj.Text() = num_obj
    msg_received = msg_received.Substring(indx + 1)
    For index = 1 To (num_obj - 1) Step 1
        indx = msg_received.IndexOf("#")
        object_cam(index) = msg_received.Substring(0, indx - 1)
        list_cam.Items.Item(index - 1) = object_cam(index)
        msg_received = msg_received.Substring(indx + 1)
    Next
    index = num_obj
    indx = msg_received.IndexOf("#")
    object_cam(index) = msg_received.Substring(0, indx - 1)
    list_cam.Items.Item(index - 1) = object_cam(index)
Else
    ans_robot_3.Text() = "no objects"
End If
sock.Close()
End If
```

In the code above, the information about the number and position of the identified objects is extracted from the returned string and listed in the list-box and other output textboxes. The user can then select one of the obtained positions and command the robot to pick that object and place it on the output container box. The code below is the implementation of the "*Pick*" action (Figure 3.25):

```
sock = ConnectSocket(ip2.Text.ToString, port2.Text.ToString)
Pick.Enabled = False
If sock Is Nothing Then
    ans_robot_3.Text() = "Error connecting, master."
Else
    Dim bytesSent As [Byte]() = Nothing
    bytesSent = ascii.GetBytes("command_str 5000_" +
                object_cam(list_cam.SelectedIndex + 1))
    sock.Send(bytesSent, bytesSent.Length, 0)
    bytes = sock.Receive(bytesReceived, bytesReceived.Length, 0)
    ans_robot.Text() = Encoding.ASCII.GetString(bytesReceived, 0, bytes)
    sock.Close()
```

```
    If Encoding.ASCII.GetString(bytesReceived, 0, bytes) = "0" Then
        ans_robot_3.Text() = "Pick command, master."
        list_cam.Items.Item(list_cam.SelectedIndex) = "no object"
    Else
        ans_robot_3.Text() = "Error executing, master."
    End If
End If
```

The "*Pick*" action is associated with a robot subroutine driven by the variable "*command_str*". The action is identified with the number 5000, and requires the user to specify also the parameters X and Y, referring to the position of the object. Consequently, the command from the client application to successfully trigger the "Pick" action is,

bytesSent = ascii.GetBytes("command_str 5000_" +
object_cam(list_cam.SelectedIndex + 1))

which translates to,

command_str 5000 X Y

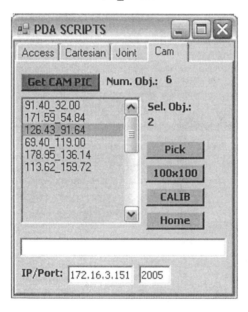

Figure 3.25 PPC screen designed to interface the Webcam

The robot subroutine handles these commands in the way presented below:

```
IF index = receive_len+1 THEN
  command_str:=receive_string1;
ENDIF
IF (index > 1) and (index < receive_len) THEN
  command_str:=StrPart(receive_string1,1,index-1);
  str_aux1:=StrPart(receive_string1,index+1,receive_len-index);
  receive_len:=StrLen(str_aux1);
  index:=StrMatch(str_aux1,1,"_");
  IF index = (receive_len + 1) THEN
    parameter1a:=str_aux1;
  ENDIF
  IF (index > 1) and (index < receive_len) THEN
    parameter1a:=StrPart(str_aux1,1,index-1);
    str_aux2:=StrPart(str_aux1,index+1,receive_len-index);
    receive_len:=StrLen(str_aux2);
    index:=StrMatch(str_aux2,1,"_");
    IF index = (receive_len + 1) THEN
      parameter2a:=str_aux2;
    ENDIF
  ENDIF
  IF (index > 1) and (index < receive_len) THEN
    parameter2a:=StrPart(str_aux2,1,index-1);
    str_aux3:=StrPart(str_aux2,index+1,receive_len-index);
    receive_len:=StrLen(str_aux3);
    index:=StrMatch(str_aux3,1,"_");
    IF index = (receive_len + 1) THEN
      parameter3a:=str_aux3;
    ENDIF
    IF (index > 1) and (index < receive_len) THEN
      parameter3a:=StrPart(str_aux3,1,index-1);
    ENDIF
  ENDIF
ENDIF
TEST command_str
  case "190": movecontact;
  case "200": open_g;
  case "201": close_g;
  case "301": move_P1;
  case "401": go_home;
  case "501": movej1p;
  case "502": movej1m;
  case "503": movej2p;
  case "504": movej2m;
  case "505": movej3p;
  case "506": movej3m;
```

```
      case "507": movej4p;
      case "508": movej4m;
      case "509": movej5p;
      case "510": movej5m;
      case "511": movej6p;
      case "512": movej6m;
      case "520": jammount1;
      case "530": cammount1;
      case "540": pick_pen;
      case "550": release_pen;
      case "1000": save_pos;
      case "2000": move_table;
      case "3000": exe_script;
      case "5000": cam_pick;
      case "5001": cam_go;
   ENDTEST
```

Basically, the routine extracts the information from the command string sent through the socket connection, and feeds the controlling variables with the commanded values. The TEST cycle (similar to a switch-case-do cycle) discriminates the function to call, which executes the functionality commanded by the user.

This example shows in some detail the procedure to explore TCP/IP socket servers for industrial manufacturing systems. It also shows that there are several platforms available to simplify the HMI and the setup, making the overall application easier to use.

3.4.4 Using UDP Datagrams

Using UDP datagrams (socket datagrams) is not fundamentally different than using TCP sockets (stream datagrams). Consequently, a simple implementation is mentioned here with the objective of pointing out the practical. The selected implementation uses a MOTOMAN robot (model HP6) equipped with the new NX100 robot controller. This controller offers remote services available from a UDP socket server, which are similar in functionality to the ones listed in Table 3.3. Several client applications were developed by the author to access those services, including the secondary services built based on those available from the UDP server, using *the Microsoft Visual Studio .NET 2005* programming suite. In the following, a simple application developed to run on Pocket PC (running *Windows Mobile 2005*) will be briefly introduced.

When using UDP datagrams, which are unreliable connections, the user should not use blocking calls, i.e., connections that block the application while waiting on the socket for the answer to the call. Consequently, after opening a socket and sending a UDP datagram, the user program shouldn't wait forever for an answer on the socket or thread. Instead, it should close the socket based on a timeout event. The following application (Figure 3.26) runs on PPC and makes a few UDP datagram calls to the UDP socket server running on the robot controller.

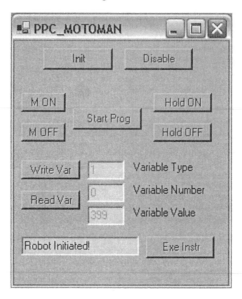

Figure 3.26 PPC application designed for a *Motoman* robot to explore UDP services from its NX100 controller

The program running on the robot controller, to implement operational (or secondary) services, is a switch-case-do type cycle driven by a numeric variable (type 1, index 0 – in the motoman notation). The simple server for this application moves the robot to five fixed positions, depending on the value of the above mentioned variable:

```
WHILE never_end
  WAIT B00 <> 0;
  TEST B00
    Case 399
      MOVE P1, VEL, 0, T0;
    Case 499
      MOVE P2, VEL, 0, T0;
    Case 599
      MOVE P3, VEL, 0, T0;
    Case 699
      MOVE P4, VEL, 0, T0;
    Case 799
      MOVE P5, VEL, 0, T0;
  ENDTEST
```

```
  B00 = 0;
RETURN
```

Writing, for example, the value 399 in the variable B00 makes the robot move to position P1. The code associated with requesting that action remotely is:

```
Dim remoteIP As New IPEndPoint(IPAddress.Parse("172.16.0.93"), 10006)
Dim Socket_send As New Socket(remoteIP.AddressFamily, SocketType.Dgram,
ProtocolType.Udp)
Dim Socket_receive As New UdpClient(10006)
Dim ENQ() As Byte = {&H6, &H0, &H1, &H0, &H5}
Dim EOT() As Byte = {&H6, &H0, &H1, &H0, &H4}
Dim ACK0() As Byte = {&H6, &H0, &H2, &H0, &H10, &H30}
Dim ACK1() As Byte = {&H6, &H0, &H2, &H0, &H10, &H31}
Socket_send.Connect(remoteIP)
Socket_receive.Connect(remoteIP)
Dim str_temp As String
Socket_send.Send(ENQ)
Dim receiveBytes As [Byte]() = Socket_receive.Receive(remoteIP)
recb = receiveBytes.Length()
For i As Integer = 0 To recb - 1
  str_temp = str_temp + Hex(receiveBytes(i))
Next i
If str_temp <> "60201030" Then
  MessageBox.Show("Erro na resposta ao ENQ: " + str_temp)
  Socket_send.Close()
  Socket_receive.Close()
  Return
End If

Dim str_temp As String
Socket_send.Send(Comando)
Dim receiveBytes As [Byte]() = Socket_receive.Receive(remoteIP)
recb = receiveBytes.Length()
For i As Integer = 0 To recb - 1
  str_temp = str_temp + Hex(receiveBytes(i))
Next i
If str_temp <> "60201031" Then
  MessageBox.Show("Erro na resposta ao comando: " + str_temp)
  Socket_send.Close()
  Socket_receive.Close()
  Return
End If
...
Send End Of Transmission
Send ACK0
Send ACK1
```

...

Socket_send.Close()
Socket_receive.Close()
This code is rather complex, since all the details about the protocol, including the negotiation phases, are explicitly programmed in the function. Basically, to send a command the protocol adopted by *Motoman* requires a command start, followed by the command itself, and then an end-of-command sequence.

The reader should remember that the sockets named "*socket_receive*" have a pre-defined timeout that prevents the application from blocking. When a timeout occurs, the routine returns immediately.

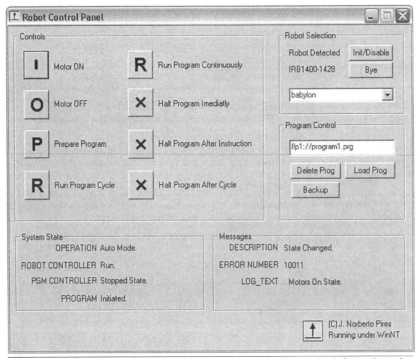

Figure 3.27 Control panel application events ("*messages*") received from the robot controller

3.5 Simple Example: Control Panel

The "*Control Panel*" is rather different from the previous examples. First, it uses *remote procedure calls* (RPCs) to access the services available from the remote server, which is a standard way to offer services and to support client-server programming environments. Other than that, the application works also as an RPC server, because it is capable of receiving events from the robot controller. The events are RPC calls made by the controller to the machines that made subscriptions to receive those events.

The application was built using PCROBNET2003/5 [5-7], an *ActiveX* software component that offers the methods, properties, and data structures necessary to explore the RPC services from the robot controller (ABB S4 robot controllers).The code for some selected actions is briefly explored below. For example, the code (developed in C++ using methods from the above mentioned *ActiveX* component) for the actions *"MOTOR ON"*, *"MOTOR OFF"*, *"PROGRAM RUN"*, and *"PROGRAM STOP"* is presented below:

```
void CCtrpanelDlg::Onmotoron()
{
  nresult = m_pon.MotorON();         ◄─────────────── Call method
  if (nresult == -8999) no_comms = TRUE;
}
```

```
void CCtrpanelDlg::Onmotoroff()
{
  nresult = m_pon.MotorOFF();        ◄─────────────── Call method
  if (nresult == -8999) no_comms = TRUE;
}
```

```
void CCtrpanelDlg::Onrunprogramcon()
{
  long cycles = -1;
  long mode = 1;
  nresult = m_pon.ProgStart("main",&cycles, &mode);  ◄─── Call method
  if (nresult == -8999) no_comms = TRUE;
}
```

```
void CCtrpanelDlg::Onhaltprogramim()
{
  short mode = 3;
  nresult = m_pon.ProgStop(&mode);   ◄─────────────── Call method
  if (nresult == -8999) no_comms = TRUE;
}
```

To receive events, a specially developed RPC server must be running on the *client* computer to receive those RPC calls. That server broadcasts the received events as registered operating system user messages (Figure 3.13). Consequently, to be able to receive those events, each application just needs to watch its message queue and filter the relevant messages. The code below was designed to operate on the message queue to identify events and present the information to the user (see *"messages"* in Figure 3.27).

```
void   CCtrpanelDlg::OnSponMsgPcroB.C.trl1(long   FAR*   msg_number,   long
FAR* msg_lParam, long FAR* msg_wParam)
{
 BSTR msg;
 m_pon.ReadMsg(&msg, msg_lParam, msg_wParam);
 CString Msg(msg);
 m_logtext.SetWindowText(Msg);
 SysFreeString(msg);
 switch (*msg_lParam)
  {
    case 1: m_description.SetWindowText("State Changed."); break;
    case 2: m_description.SetWindowText("Warning."); break;
    case 3: m_description.SetWindowText("Error."); break;
    default: m_description.SetWindowText("Invalid log_type."); break;
  }
 Msg.Format("%d",*msg_wParam);
 m_error.SetWindowText(Msg);
 CCtrpanelDlg::info();
}
```

Using software components (*ActiveX*, *JAVA*, etc.) is a way to hide from the user
the tricky details about how to make RPC calls (for example, compare this code
with the one presented for the UDP datagram example), allowing her to focus
immediately on the application.

3.6 Simple Example: S4Misc – Data Access on a Remote Controller

The "*S4Misc*" application (Figure 3.28) also uses RPC to access the robot services.
Like the previous example, it was designed to be used with the ABB S4 robot
controllers (running option RAP [8]).

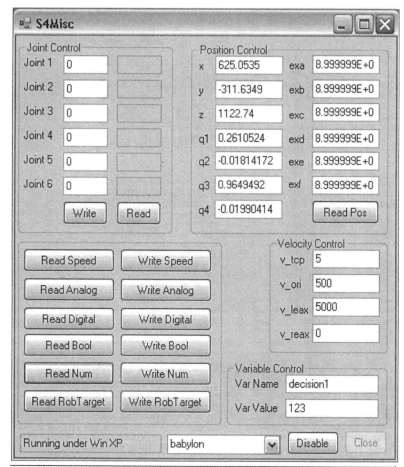

Figure 3.28 *S4Misc* application designed to access program and system variables from a remote computer

This application enables the user to access program and system variables from a remote computer online, i.e., even when the robot is in automatic mode and the loaded program is executing. The user can utilize this software for debugging purposes, checking and changing (when needed) the actual value of any variable. In the following, the code for the actions *READ/WRITE* a numeric variable, *WRITE* a speed variable, and *READ* the actual robot position is showed (*C# .Net 2005* was used here):

```
private void OnReaNum()
{
    String msg;
    msg = txt_VarName.Text;
    if (msg.Length > 0)
```

```
    {
       nresult = PcRob.ReadNum(msg, ref val);   ◄——— Call method
       if (nresult < 0)
       {
          MessageBox.Show("Error Reading Num!");
       }
       else
       {
          msg = Convert.ToString(val);
          txt_VarValue.Text = msg;
       }
    }
    else MessageBox.Show("Error: You must specify variable name!");
}

private void OnWriteNum()
{
    String msg;
    String msg1;
    msg = txt_VarName.Text;
    msg1 = txt_VarValue.Text;
    if (msg.Length > 0 || msg1.Length > 0)
    {
       val = Convert.ToSingle(msg1);
       nresult = PcRob.WriteNum(msg, ref val);   ◄——— Call method
       if (nresult < 0) MessageBox.Show("Error Wrinting Num!");
    }
    else MessageBox.Show("Error: You must specify variable name and value!");
}


private void OnWriteSpeed()
{
    String msg;
    msg = txt_VarName.Text;
    if (msg.Length > 0)
    {
       RobVelocity.vtcp = Convert.ToSingle(txt_VTcp.Text);
       RobVelocity.vori = Convert.ToSingle(txt_VOri.Text);
       RobVelocity.vleax = Convert.ToSingle(txt_VLeax.Text);
       RobVelocity.vreax = Convert.ToSingle(txt_VReax.Text);
       PcRob.vtcp = RobVelocity.vtcp;
       PcRob.vori = RobVelocity.vori;
       PcRob.vleax = RobVelocity.vleax;
       PcRob.vreax = RobVelocity.vreax;
       nresult = PcRob.WriteSpeedDataVB(msg);   ◄——— Call method
       if (nresult<0) MessageBox.Show("Error: You must specify variable name");
    }
```

```
    else MessageBox.Show("Error: You must specify variable name");
}

private void OnReadCurrRoboTarget()
{
    nresult =PcRob.ReadCurrRobTVB();   ◄——— Call method
    if (nresult < 0)
    {
        MessageBox.Show("Error Reading Current RobT");
    } else
    {
        RobT_Read.x = PcRob.x;
        RobT_Read.y = PcRob.y;
        RobT_Read.z = PcRob.z;
        RobT_Read.q1 = PcRob.q1;
        RobT_Read.q2 = PcRob.q2;
        RobT_Read.q3 = PcRob.q3;
        RobT_Read.q4 = PcRob.q4;
        RobT_Read.exa = PcRob.exa;
        RobT_Read.exb = PcRob.exb;
        RobT_Read.exc = PcRob.exc;
        RobT_Read.exd = PcRob.exd;
        RobT_Read.exe = PcRob.exe;
        RobT_Read.exf = PcRob.exf;
        txt_x.Text = RobT_Read.x.ToString();
        txt_y.Text = RobT_Read.y.ToString();
        txt_z.Text = RobT_Read.z.ToString();
        txt_q1.Text = RobT_Read.q1.ToString();
        txt_q2.Text = RobT_Read.q2.ToString();
        txt_q3.Text = RobT_Read.q3.ToString();
        txt_q4.Text = RobT_Read.q4.ToString();
        txt_exa.Text = RobT_Read.exa.ToString();
        txt_exb.Text = RobT_Read.exb.ToString();
        txt_exc.Text = RobT_Read.exc.ToString();
        txt_exd.Text = RobT_Read.exd.ToString();
        txt_exe.Text = RobT_Read.exe.ToString();
        txt_exf.Text = RobT_Read.exf.ToString();
    }
}
```

This application demonstrates the usefulness of having remote services that can communicate with the running applications. With it, users can influence the behavior of running applications for controlling, monitoring, or debugging purposes. It also demonstrates the usefulness of software components for the process of developing distributed applications that necessarily use several types of radically different equipment. With these components, users and programmers can focus on the applications under development without worrying about the technical details of remote procedure calling, network communications, and so on.

3.7 Industrial Example: Semi-autonomous Labeling System

In this section, an industrial example that explores the previous material is presented and discussed. This example corresponds to an actual implementation resulting from a cooperation effort between the author and a Portuguese company. The system presented here was designed to operate almost without operator intervention, showing that concepts like flexibility and agility are fundamental to manufacturing plants and require much more from the systems used on the shop floor. Flexible manufacturing systems take advantage of being composed of programmable equipment to implement most of its characteristics, which are supported by reconfigurable mechanical parts. Industrial robots are, consequently, good examples of flexible manufacturing systems.

The robotic industrial system presented here was designed to execute parameterized labeling tasks on paper rolls. The system is commanded directly from the manufacturing tracking and control software. This software is based on dynamic databases that register the situation of each item produced in the factory, a simple way to track them see what is happening on the shop floor. Since all information about each item is available in the manufacturing tracking software, it is logical to use it to command some of the shop floor manufacturing systems, namely the ones that require only simple parameterization to work properly. This procedure would take advantage of the available information and infrastructure, avoiding unnecessary operator interfaces to command the system. Also, potential gains in terms of flexibility and productivity are evident.

Figure 3.29 Labeling system

3.7.1 Robotic Labeling System

The industrial system introduced here is a labeling system (Figure 3.29) composed of:

- One robot manipulator ABB IRB4400, with the S4C+ controller [10]
- One electro-pneumatic gripper, properly equipped to grab one or two A4-size paper sheets
- One office laser printer, with several trays of paper
- One gluing machine with spray injectors controlled from the robot controller IO system
- One industrial PLC (Siemens S7-300) that controls the rolls conveyer belt, providing information to the robot controller about its state

In general, the labeling robotic system works as follows: When a roll is released from the previous system (wrapping machine), one or two labels are printed on the laser printer. At the same time, the robot receives the order to pick those labels from the ramp placed at the end of the printer, and immediately prepositions near the printer. The picking operation happens when the required number of sheets are available at the ramp (two optical sensors detect the presence of paper). After that, the robot waits for the roll to enter the working zone, i.e., waits for the corresponding optical sensor, named sensor 1 in Figure 3.29, to detect the roll. When the roll is detected, the robot moves to the gluing machine to add glue on the side of each label. When the operation is finished, the roll should be already stopped, waiting for the robot to insert the labels on the top and on the right side of the roll. The robot performs that operation when the roll is detected by sensor 2 (Figure 3.29) and when the PLC confirms that the conveyor has stopped. When the operation is finished, the robot signals it using a flag, accessible remotely, and moves to a neutral position to wait for a new command.

3.7.2 System Software

Designing software for the system, which needed to be commanded from the network, was an interesting challenge. The industrial robot is the central element of the manufacturing cell, and is connected to the factory network, which makes it easily accessible from the UNIX station running the manufacturing tracking software.

To exchange information between computer systems, in a safe and guaranteed way, a client-server approach using TCP/IP sockets may be used. That is a simple and straightforward thing to do, with the UNIX computer acting as the client. A TCP/IP server should then be available to receive client calls, and a properly designed messaging protocol must be used. The decision here was to make the TCP/IP server the only interface to the robotic manufacturing cell, so that any command or request of information is done by connecting to the server and sending the appropriate messages. Since there is a network on the shop floor, the TCP/IP server can be installed in any shop floor computer, making it really easy to install the interface and have it running. In the factory under consideration, the majority of the shop floor computers are running the *Windows NT4* and *Windows 2000* operating systems. Consequently, we decided to use BSD compatible TCP/IP sockets, which are also compatible with the Microsoft TCP/IP implementation (*winsock2*).

The next challenge was how to manage the communication with the robot controller, since it is well known that actual robot controllers are closed industrial systems not allowing installation of any user software apart from robot programs. ABB robot controllers [10] have internal *Remote Procedure Call* (RPC) [2,8] servers that can be used to exchange variables, files, etc. Those servers are *SUN RPC 4.0* [2] compatible, and can be used to our purposes if the TCP/IP server interface can issue RPC calls to the robot controller. Consequently, a library of functions implementing calls for all the services on the ABB robot controller was built [5,7], along with a port of the *SUN RPC 4.0* to operating systems based on the *Win32* API. This environment enables a complete access to the robot controller RPC services making it possible to command the robot from the network. The robot controller software must then be built in a way to expose all system capabilities to the remote client. This means building it like a *SWITCH-CASE-DO* server, with the switching variable controlled by the remote client.

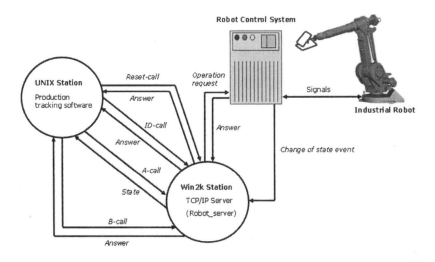

Figure 3.30 Software interface to the industrial robotic system

The basic idea, depicted in Figure 3.30, is simple. The interface to the industrial robotic cell is a TCP/IP server running on a specified IP and port number. The following procedure is used in a way to guarantee safety and avoid data loss:

- The server should respond to *ID-calls* with a pre-determined string, which is used to identify the TCP/IP server with name, version, and date. The string is actually "*robot_server@v21m11y03*". The *ID-call* is the first call issued by the client after establishing a new connection. A wrong answer to the *ID-call* should tell the client to send a *reset-call* and close the connection

- The client makes frequent *A-calls*, in periods of two seconds, to find out if the server is alive and healthy, and to get its actual state (*busy* or *ready*)

- The client uses *B-calls* to send execution commands, properly parameterized, to the robotic labeling system. When a *B-call* is received and accepted by the server, the system enters the *busy* state and any subsequent *A-call* will return that the system is *busy*

- When the robotic labeling system completes a task, i.e., when it inserts the requested number of labels on the roll in use, the system enters the *ready* state and any subsequent *A-call* will return that state

The TCP/IP server is the only operational interface to the robotic system. Basically, it is a simple single channel TCP/IP server, completely coded in C++, which waits for connections on a pre-determined port, accepting only the ones coming from only a few IPs (the ones where the manufacturing tracking software may be running). Connection is established only if the calling machine makes an *ID-call*, properly parameterized, including a password. The server is a state machine that implements answers to the four different messages that can be sent by the connected client (Figure 3.30). The connection between the TCP/IP server and the industrial robot is handled using RPC sockets, compatible with the SUN RPC 4.0 definition.

In the following section, the developed software will be further explained, starting with the software designed to run on the robot controller.

3.7.3 Robot Controller Software

Considering that the system was designed to be commanded remotely using the factory computer network, it was decided to have the robot controller software working as a server, exposing to the remote client all of its operational functionalities. This capability is very interesting also for other applications, and because of that it will be discussed in a general way.

When building an industrial robotic cell, it is certainly possible, and very useful, to identify all the system capabilities and requirements, i.e., the system engineer should state clearly all the functions it is supposed to perform and write the code necessary to implement them. If that code is developed as general as possible, and used to build a server that can be explored remotely with properly parameterized calls, then the complete system functionality can be requested remotely from the network.

Technically, to implement the remote calls, it was decided to *use remote procedure calls* (RPC) compatible with the *SUN RPC 4.0* suite, an open standard in the public domain. The ABB S4 robot control system implements a collection of RPC services that enable users to access programs, system data, and robot configuration, as well as to share files, etc. These services are part of the robot controller's operating system. Using those services from the TCP/IP server designed to interface the system [2, 5-8], it is certainly possible to set up an RPC-driven server like:

```
switch service_decision_variable
    case 1: call function_1; break;
    case 2: call function_2; break;
    case 3: call function_3; break;
    ...
    case n: call function_n; break;
    default: call invalid_function; break;
end_switch
```

where the *service_decision_variable* is a numerical variable whose value can be changed remotely, making an RPC call to the *change_numerical_value* service. In this way, the robot's operation is completely controlled from the remote client.

Since the robotic system is to be operated without human intervention, a few services were added to allow maintenance and error recovery operations. Sometimes, due to errors in the manufacturing tracking database (usually introduced by human intervention), invalid or badly parameterized commands are sent to the robot. In those situations, depending on the dimensions of the roll in use, the robot may crash with the surface of the roll, because it uses the commanded dimensions to approach the surface of the roll in a more efficient way. Also, failure in the conveyor sensors or actuators may cause problems with roll placement. In any case, an operator is required to solve the problems and put the system in production again. The program shown in Figure 3.31 is used to put the robot in a known position and resume automatic operation.

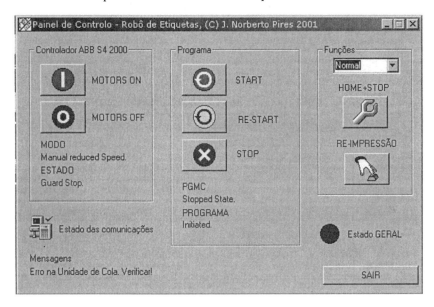

Figure 3.31 Operator interface used to solve error situations

Basically, the operator interface makes RPC calls to the above mentioned services designed to solve erroneous situations. Those services enable the operator to resume local program execution from an actual point or from the beginning, move the robot to well-known positions, enter maintenance routines, and so on. The program usually runs on a laptop that maintenance personnel carry to the setup when a problem arises, plugging it to the network.

3.7.4 TCP/IP Server

As already explained, this TCP/IP server (Figure 3.32) was developed as the only interface to the robotic labeling system. It is a simple TCP/IP server that accepts connections coming from the machine that runs the manufacturing tracking software (client). After connecting, it implements a state-machine that listens for messages coming from the client, acting accordingly. The TCP/IP server monitors the connection to the robot and the robot state, so that proper answers are given to every *A-call* received from the client. Also, the server does not accept any command in the periods where the robot state is *busy*, forcing the client to wait until the previous commanded operation finishes.

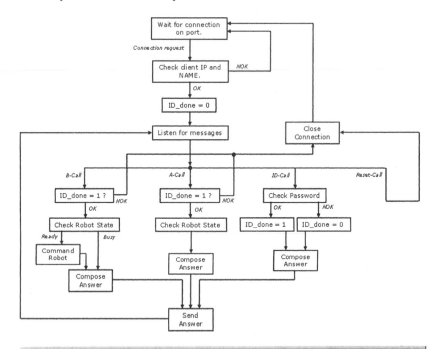

Figure 3.32 TCP/IP server operation

3.7.5 Discussion

The example presented in this section explores the use of software interfaces for remote command of shop floor industrial robotic cells. This involves network interfaces based on the TCP/IP protocol and remote procedure calls, enabling direct command of shop floor manufacturing setups from anywhere in the factory. This idea is particularly useful with systems that require minor parameterization to perform a predefined task. This can easily be done from the manufacturing tracking software, used to follow and manage production, where the require information is available.

In many industries, like the one presented in this example, production is closely tracked in any part of the manufacturing cycle. The manufacturing cycle can be interpreted as a collection of operations and processes that transform the raw materials into finished products. This journey of products between the raw materials warehouse and the finished products warehouse, is composed of several manufacturing systems that perform the necessary operations on the product under processing, and intermediate buffers used to temporarily store semi-finished products in several stages of their production cycle. These buffers are fundamental for a well balanced production planning, achieving high rates of efficiency and agility. In many cases, the manufacturing systems require only minor parameterization to execute their tasks. If that parameterization can be commanded remotely from where it is available (using manufacturing tracking software), then the system becomes almost autonomous in the sense that operator intervention is minimal and related only with maintenance and error situations. A system like this will improve manufacturing efficiency and agility, since the operation becomes less dependent on operators. Also, because the system was built to be explored remotely, which requires a collection of general software routines designed to implement all of the system functionalities, it is easier to change production by changing parameterization, a software task, which also contributes to agility.

This robotic manufacturing system uses a simple TCP/IP server as the commanding interface. The server sends remote procedure calls to the robot control system, which is the system main computer. The robot controller interfaces with the system PLC that controls the conveyor, and manages the information coming from manual controls and sensors. Consequently, any client connected to the TCP/IP server is able to command the system and get production information. This feature adds flexibility and agility to the manufacturing setup. This setup was installed in a Portuguese paper factory and is being used without problems for almost three years, which demonstrates its robustness and simplicity.

Finally it is worthwhile to stress that:
- The system interface was implemented in C++ using available programming tools: *Visual C++ 6.0* first, and *Visual .NET 2003* when an update was necessary [11]

- The system was implemented using standard operating systems, namely, *UNIX* from *Digital* to run the manufacturing tracking software, and *Windows 2000* to run the robotic cell TCP/IP interface
- The *Microsoft* TCP/IP socket implementation was used to program the TCP/IP server, since it is BSD-compatible
- The system uses RPC's compatible with *SUN RPC 4.0*, an open standard not compatible with the *Microsoft* implementation, which required a complete port to *Windows 2000* (the operating system used on the shop floor of the partner factory). That effort was completely done by the author

Consequently, no special tools were used to build the presented solution, which proves that these techniques are available to build smart interfaces enabling more efficient applications, or at least to build other ways to exploit shop floor equipment. In fact, actual manufacturing systems have a lot of flexibility inside because they rely on programmable equipment, like robots and PLCs, to implement their functions. System engineers need to find ways to explore that flexibility when designing manufacturing systems, exposing it to the advanced user in more efficient ways.

In terms of operational results, it is important that a system like the one presented here does not add any production delay to the manufacturing system, or become a production bottleneck. This means that the cycle time should be lower than the cycle time of the previous station. In our example, the system takes around 11 seconds to perform the labeling operation, which is at least 20 seconds lower than the previous roll wrapping operation.

3.7.6 Conclusion

In describing an industrial application designed for labeling applications, this section discussed and detailed a software interface designed to command shop floor manufacturing systems remotely from the manufacturing tracking software. This interface added flexibility and agility to the manufacturing system, since all available operations were implemented in a very general way requiring only simple parameterization to specify the individual operations. The interface to the system is a simple TCP/IP server installed in one of the shop floor computers. To command the system, the client needs to connect to the server and, if allowed, send properly parameterized commands as simple messages over the open socket. The server uses *SUN RPC 4.0* compatible sockets to access the robotic system, place the received commands, and monitor the system operation. Since the TCP/IP server is a general implementation, using the BSD compatible TCP/IP socket implementation from *Microsoft*, it can receive commands from virtually any client. This makes the presented robotic cell interface an interesting way to command shop floor manufacturing systems.

3.8 References

[1] Halsall F., "Data Communications, Computer Networks and Open Systems", Third Edition, Addison-Wesley, 1992.

[2] Bloomer J., "Power Programming with RPC", O'Reilly & Associates, Inc., 1992.

[3] Siemens, "S7-200 System and Programming Manual", Siemens Automation, 2003.

[4] Mahalik, NP, "FieldBus Technology, Industrial Network Standards for Real-Time Distributed Control", Springer, 2003.

[5] Pires, JN, and Loureiro, Altino et al, "Welding Robots", IEEE Robotics and Automation Magazine, June, 2003

[6] Pires, JN, "Using Matlab to Interface Industrial Robotic & Automation Equipment", IEEE Robotics and Automation Magazine, September 2000

[7] Pires, JN, Sá da Costa, JMG, "Object-Oriented and Distributed Approach for Programming Robotic Manufacturing Cells", IFAC Journal Robotics and Computer Integrated Manufacturing, Volume 16, Number 1, pp. 29-42, March 2000.

[8] ABB Robotics, "RAP Service Specification, ABB Robotics, 2000.

[9] ABB Robotics, "S4C+ documentation CD" and "IRC5 documentation CD", ABB Robotics, 2000 and 2005

[10] ABB IRB140, IRB1400, IRB1400 & IRB6400 System Manual, ABB Robotics, Vasteras, Sweden, 2005.

[11] Visual Studio.NET 2003 Programmers Reference, Microsoft, 2003 (reference can be found at Microsoft's web site in the Visual C++ .NET location)

[12] Visual Studio.NET 2005 Programmers Reference, Microsoft, 2005 (reference can be found at Microsoft's web site in the Visual C++ .NET location)

4

Interface Devices and Systems

4.1 Introduction

The success of using robots with flexible manufacturing systems especially designed for small and medium enterprises (SME) depends on the human-machine interfaces (HMI) and on the operator skills. In fact, although many of these manufacturing systems are semi-autonomous, requiring only minor parameterization to work, many other systems working in SMEs require heavy parameterization and reconfiguration to adapt to the type of production that changes drastically with time and product models. Another difficulty is the average skill of the available operators, who usually have difficulty adapting to robotic and/or computer-controlled, flexible manufacturing systems.

SMEs are special types of companies. In dimension (with up to 250 permanent collaborators), in economic strength (with net sales up to 50M€) and in installed technical expertise (not many engineers). Nevertheless, the European economy depends on these types of company units since roughly they represent 95% of the European companies, more than 75% of the employment, and more than 60% of the overall net sales [1]. This reality configures a scenario in which flexible automation, and robotics in particular, play a special and unique role requiring manufacturing cells to be easily used by regular non-skilled operators, and easier to program, control and monitor. One way to this end is the exploitation of the consumer market's input-output devices to operate with industrial robotic equipment. With this approach, developers can benefit from the availability, and functionality of these devices, and from the powerful programming packages available for the most common desktop and embedded platforms. On the other hand, users could benefit from the operational gains obtained by having the normal tasks performed using common devices, and also from the reduction in prices due to the use of consumer products.

Industrial manufacturing systems would benefit greatly from improved interaction devices for human-machine interface even if the technology is not so advanced. Gains in autonomy, efficiency, and agility would be evident. The modern world requires better products at lower prices, requiring even more efficient manufacturing plants because the focus is on achieving better quality products, using faster and cheaper procedures. This means having systems that require less operator intervention to work normally, better human-machine interfaces, and cooperation between humans and machines sharing the same workspace as real coworkers.

Also, the robot and robotic cell programming task would benefit very much from improved and easy-to-use interaction devices. This means that availability of SDKs and programming libraries supported under common programming environments is necessary. Application development depends on that.

Working on future SMEs means considering humans and machines as coworkers, in environments where humans have constant access to the manufacturing equipment and related control systems.

Several devices are available for the user interface (several types of mice, joysticks, gamepads and controls, digital pens, pocket PCs and personal assistants, cameras, different types of sensors, etc.) with very nice characteristics that make them good candidates for industrial use. Integrating these devices with current industrial equipment requires the development of a device interface, which exhibits some basic principles in terms of software, hardware and interface to commercial controllers.

This scenario can be optimized in the following concurrent ways:

1. Develop user-friendly and highly graphical HMI applications to run on the available interface devices. Those environments tend to hide the complexity of the system from operators, allowing them to focus on controlling and operating the system. Figure 4.1 shows the main window of an application used to analyze force/torque data coming from a robotic system that uses a force/torque sensor to adjust the programmed trajectories (this system will not be further explored in this book)

2. Explore the utilization of consumer input/output devices that could be used to facilitate operator access to the system. In fact, there is a considerable amount of different devices on the market developed for personal computers on different input/output tasks. Such devices are usually programmable, with the manufacturers providing suitable SDKs to make them suitable for integrating with industrial manufacturing systems. Figure 4.2 shows a few of these devices, some of them covered in this book

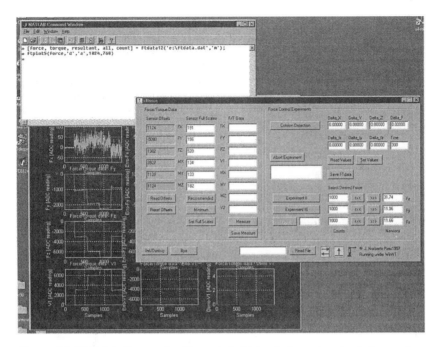

Figure 4.1 HMI interface used with an industrial robotic system to further analyze force/torque sensor data

3   Explore the functionality of the available software packages commonly used for engineering. Good examples of those packages are the CAD packages used by engineers to develop, optimize, and improve their designs (Figure 4.3). Since the vast majority of companies use CAD software packages to design their products, it would be very interesting if the information from CAD files could be used to generate robot programs. That is, the CAD application could be the environment used for specifying how robots should execute the required operations on the specified parts. Furthermore, since most engineers are familiar with CAD packages, exploring CAD data for robot programming and parameterization seems a good way to proceed [2].

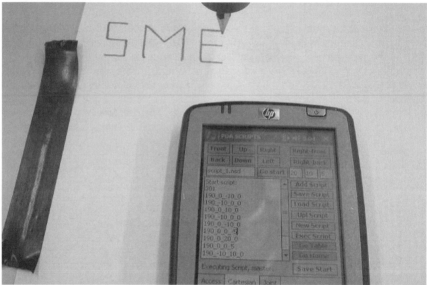

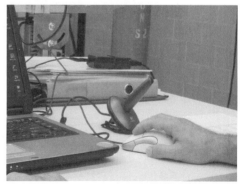

Figure 4.2 Input/output devices used for HMI applications: (from top to bottom) joystick, headset with noise reduction, pocket PC and digital pen

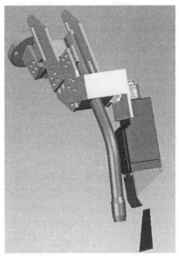

a)

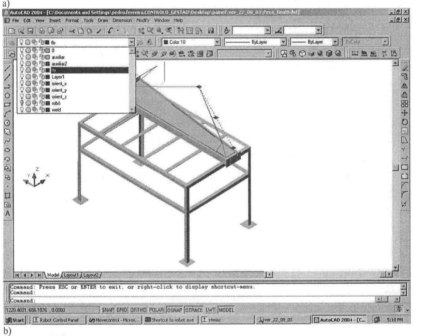

b)

Figure 4.3 Using 3D CAD software packages to project and design mechanical parts: a – welding torch and laser camera (*SolidWorks*); b – welding trajectories specified using *AutoCad*

This chapter uses industrial and laboratory test-cases to provide the necessary details and insight to complement the above presented claims and design options.

4.2 Speech Interfaces

4.2.1 Introduction

Talking to machines is a thing normally associated with science fiction movies and cartoons and less with current industrial manufacturing systems. In fact, most of the papers about speech recognition start with something related to artificial intelligence, a science fiction movie, or a robot used in a movie, etc., where machines talk like humans, and understand the complex human speech without problems. Nevertheless, industrial manufacturing systems would benefit very much from speech recognition for human-machine interface (HMI) even if the technology is not so advanced. Gains in terms of autonomy, efficiency and agility seem evident. The modern world requires better products at lower prices, requiring even more efficient manufacturing plants because the focus is in achieving better quality products, using faster and cheaper procedures. This means autonomy, having systems that require less operator intervention to operate normally, better human-machine interfaces and cooperation between humans and machines sharing the same workspace as real coworkers.

The final objective is to achieve, in some cases, semi-autonomous systems [3], i.e., highly automated systems that require only minor operator intervention. In many industries, production is closed tracked in any part of the manufacturing cycle, which is composed by several in-line manufacturing systems that perform the necessary operations, transforming the raw materials in a final product. In many cases, if properly designed, those individual manufacturing systems require simple parameterization to execute the tasks they are designed to execute. If that parameterization can be commanded remotely by automatic means from where it is available, then the system becomes almost autonomous in the sense that operator intervention is reduced to the minimum and essentially related with small adjustments, error and maintenance situations [3]. In other cases, a close cooperation between humans and machines is desirable although very difficult to achieve, due to limitations of the actual robotic and automation systems.

The above described scenario puts focus on HMI, where speech interfaces play an important role because manufacturing system efficiency will increase if the interface is more natural or similar to how humans command things. Nevertheless, speech recognition is not a common feature among industrial applications, because:

- The speech recognition and text-to-speech technologies are relatively new, although they are already robust enough to be used with industrial applications
- The industrial environment is very noisy which puts enormous strain on automatic speech recognition systems
- Industrial systems weren't designed to incorporate these types of features, and usually don't have powerful computers dedicated to HMI

Automatic speech recognition (ASR) is commonly described as converting speech to text. The reverse process, in which text is converted to speech (TTS), is known as *speech synthesis*. Speech synthesizers often produce results that are not very natural sounding. Speech synthesis is different from voice processing, which involves digitizing, compressing (not always), recording, and then playing back snippets of speech. Voice processing results are natural sounding, but the technology is limited in flexibility and needs more disk storage space compared to speech synthesis.

Speech recognition developers are still searching for the perfect human-machine interface, a recognition engine that understands any speaker, interprets natural speech patterns, remains impervious to background noise, and has an infinite vocabulary with contextual understanding. However, practical product designers, OEMs, and VARs can indeed use today's speech recognition engines to make major improvements to today's markets and applications. Selecting such an engine for any product requires understanding how the speech technologies impact performance and cost factors, and how these factors fit in with the intended application.

Using speech interfaces is a big improvement to HMI systems, because of the following reasons:

- Speech is a natural interface, similar to the "*interface*" we share with other humans, that is robust enough to be used with demanding applications. That will change drastically the way humans interface with machines
- Speech makes robot control and supervision possible from simple multi-robot interfaces. In the presented cases, common PCs were used, along with a normal noise-suppressing headset microphone
- Speech reduces the amount and complexity of different HMI interfaces, usually developed for each application. Since a PC platform is used, which carry currently very good computing power, ASR systems become affordable and simple to use

In this section, an automatic speech recognition system is selected and used for the purpose of commanding a generic industrial manufacturing cell. The concepts are explained in detail and two test case examples are presented in a way to show that if certain measures are taken, ASR can be used with great success even with industrial applications. Noise is still a problem, but using a short command structure with a specific word as pre-command string it is possible to enormously reduce the noise effects. The system presented here uses this strategy and was tested with a simple noiseless pick-and-place example, but also with a simple welding application in which considerable noise is present.

4.2.2 Evolution

As already mentioned, the next level is to combine ASR with natural language understanding, i.e., making machines understand our complex language, coping with the implementations, and providing contextual understanding. That capability would make robots accessible to people who don't want to learn the technical details of using them. And that is really the aim, since a common operator does not have the time or the immediate interest to dig into technical details, which is, in fact, neither required nor an advantage.

Speech recognition has been integrated in several products currently available:

- Telephony applications
- Embedded systems (Telephone voice dialing system, car kits, PDAs, home automation systems, general use electronic appliances, etc.)
- Multimedia applications, like language learning tools
- Service robotics

Speech recognition has about 75 years of development. Mechanical devices to achieve speech synthesis were first devised in the early 19th century, but imagined and conceived for fiction stories much earlier.

The idea of an artificial speaker is very old, an aspect of the human long-standing fascination with humanoid *automata*. *Gerbert* (d. 1003), *Albertus Magnus* (1198-1280), and *Roger Bacon* (1214-1294) are all said to have built speaking heads. However, historically attested speech synthesis begins with *Wolfgang von Kempelen* (1734-1804), who published his findings of twenty years of research in 1791. *Wolfgang* ideas gain another interest with the invention of the telephone in the late 19th century, and the subsequent efforts to reduce the bandwidth requirements of transmitting voice.

On March 10, 1876, the telephone was born when *Alexander Graham Bell* called to his assistant, *"Mr. Watson! Come here! I want you!"* He was not simply making the first phone call. He was creating a revolution in communications and commerce. It started an era of instantaneous information-sharing across towns and continents (on a planetary level) and greatly accelerated economic development.

In 1922, a sound-activated toy dog named *"Rex"* (from *Elmwood Button Co.*) could be called by name from his doghouse.

In 1936, *U.K. Tel* introduced a *"speaking clock"* to tell time. In the 1930s, the telephone engineers at *Bell Labs* developed the famous *Voder*, a speech synthesizer that was unveiled to the public at the 1939 World's Fair, but that required a skilled human operator to operate with it.

Small vocabulary recognition was demonstrated for digits over the telephone by *Bell Labs* in 1952. The system used a very simple frequency splitter to generate plots of the first two formants. The identification was achieved by matching them with a pre-stored pattern. With training, the recognition accuracy of spoken digits was 97%.

Fully automatic speech synthesis came in the early 1960s, with the invention of new automatic coding schemes, such as *adaptive predictive coding* (APC). With those new techniques in hand, the *Bell Labs* engineers again turned their attention to speech synthesis. By the late 1960s, they had developed a system for internal use in the telephone system, a machine that read wiring instructions to *Western Electric* telephone wirers, who could then keep eyes and hands on their work.

At the *Seattle World's Fair* in 1962, IBM demonstrated the "*Shoebox*" speech recognizer. The recognizer was able to understand 16 words (digits plus command/control words) interfaced with a mechanical calculator for performing arithmetic computations by voice. Based on mathematical modeling and optimization techniques learned at IDA (now the *Center for Communications Research*, Princeton), *Jim Baker* introduced stochastic processing with *hidden markov models* (HMM) to speech recognition while at *Carnegie-Mellon University* in 1972. At the same time, *Fred Jelinek*, coming from a background of information theory, independently developed HMM techniques for speech recognition at IBM. HMM provides a powerful mathematical tool for finding the invariant information in the speech signal. Over the next 10-15 years, as other laboratories gradually tested, understood, and applied this methodology, it became the dominant speech recognition methodology. Recent performance improvements have been achieved through the incorporation of discriminative training (at *Cambridge University*, LIMSI, etc.) and large databases for training.

Starting in the 1970s, government funding agencies throughout the world (e.g. *Alvey*, *ATR*, *DARPA*, *Esprit*, etc.) began making a major impact on expanding and directing speech technology for strategic purposes. These efforts have resulted in significant advances, especially for speech recognition, and have created large widely-available databases in many languages while fostering rigorous comparative testing and evaluation methodologies.

In the mid-1970s, small vocabulary commercial recognizers utilizing expensive custom hardware were introduced by *Threshold Technology* and *NEC*, primarily for hands-free industrial applications. In the late 1970s, *Verbex* (division *of Exxon Enterprises*), also using custom special-purpose hardware systems, was commercializing small vocabulary applications over the telephone, primarily for telephone toll management and financial services (e.g. Fidelity fund inquiries). By the mid-1990s, as computers became progressively more powerful, even large vocabulary speech recognition applications progressed from requiring hardware assists to being mainly based on software. As performance and capabilities increased, prices dropped.

Further progress led to the introduction, in 1976, of the *Kurzweil Reading Machine,* which, for the first time allowed the blind to "*read*" plain text as opposed to *Braille*. By 1978, the technology was so well established and inexpensive to produce that it could be introduced in a toy, *Texas Instruments Speak-and-Spell*. Consequently, the development of this important technology from inception until fruition took about 15 years, involved practitioners from various disciplines, and had a far-reaching impact on other technologies and, through them, society as a whole.

Although existing for at least as long as speech synthesis, *automatic speech recognition* (ASR) has a shorter history. It needed much more the developments of *digital signal processing* (DSP) theory and techniques of the 1960s, such as *adaptive predictive coding* (APC), to even come under consideration for development.

Work in the early 1970s was again driven by the telephone industry, which hoped for both voice-activated dialing and also for security procedures based on voice recognition. Through gradual development in the 1980s and into the 1990s, error rates in both these areas were brought down to the point where the technologies could be commercialized.

In 1990, *Dragon Systems* (created by *Jim* and *Janet Bailer*) introduced a general-purpose discrete dictation system (i.e. requiring pauses between each spoken word), and in 1997, *Dragon* started shipping general purpose continuous speech dictation systems to allow any user to speak naturally to their computer instead of, or in addition to, typing. *IBM* rapidly followed the developments, as did *Lernout & Hauspie* (using technology acquired *from Kurzweil Applied Intelligence*), *Philips*, and more recently, *Microsoft*. Medical reporting and legal dictation are two of the largest market segments for ASR technology. Although intended for use by typical PC users, this technology has proven especially valuable to disabled and physically impaired users, including many who suffer from *repetitive stress injuries* (RSI). Robotics is also a very promising area.

AT&T introduced its automated operator system in 1992. In 1996, the company *Nuance* supplied recognition technology to allow customers of *Charles Schwab* to get stock quotes and to engage in financial transactions over the telephone. Similar recognition applications were also supplied by *SpeechWorks*. Today, it is possible to book airline reservations with *British Airways*, make a train reservation for *Amtrak*, and obtain weather forecasts and telephone directory information, all by using speech recognition technology. In 1997, *Apple Computer* introduced software for taking voice dictation in Mandarin Chinese.

Other important speech technologies include speaker verification/identification and spoken language learning for both literacy and interactive foreign language instruction. For information search and retrieval applications (e.g. audio mining) by voice, large vocabulary recognition preprocessing has proven highly effective, preserving acoustic as well as statistical semantic/syntactic information. This approach also has broad applications for speaker identification, language identification, and so on.

Today, 65 years after the *Voder* and just 45 years after APC, both ASR and TTS technologies can be said to be fully operational, in a case where a very convoluted technological history has had a modest and more or less anticipated social impact.

4.2.3 Technology

Speech recognition systems can be separated into several different classes depending on the types of utterances they have the ability to recognize. These classes are based on the fact that one of the difficulties of ASR is the ability to determine when a speaker starts and finishes an utterance. Most packages can fit into more than one class, depending on which mode they're using.

Isolated words: Isolated word recognizers usually require each utterance to have quiet (lack of an audio signal) on both sides of the sample window. It doesn't mean that it accepts single words, but does require a single utterance at a time. Often, these systems have "*listen/not-listen*" states, where they require the speaker to wait between utterances (usually doing processing during the pauses). Isolated utterance might be a better name for this class.

Connected words: Connected word systems (or more correctly "connected utterances") are similar to isolated words, but allow separate utterances to be run-together with a minimal pause between them.

Continuous speech: Continuous recognition is the next step. Recognizers with continuous speech capabilities are some of the most difficult to create because they must utilize special methods to determine utterance boundaries. Continuous speech recognizers allow users to speak almost naturally, while the computer determines the content. Basically, it's computer dictation and commanding.

Spontaneous speech: There appears to be a variety of definitions for what spontaneous speech actually is. At a basic level, it can be thought of as speech that is natural sounding and not rehearsed. An ASR system with spontaneous speech ability should be able to handle a variety of natural speech features such as words being run together, pauses, "ums" and "ahs", slight stutters, etc.

Voice verification/identification: Some ASR systems have the ability to identify specific users. This book doesn't cover verification or security systems, because user validation is done using other means.

Speech recognition, or speech-to-text, involves capturing and digitizing the sound waves, converting them to basic language units or phonemes, constructing words from phonemes, and contextually analyzing the words to ensure correct spelling for words that sound alike (such as "*write*" and "*right*").

Recognizers (also referred to as speech recognition engines) are the software drivers that convert the acoustic signal to a digital signal and deliver recognized speech as text to the application. Most recognizers support continuous speech, meaning the user can speak naturally into a microphone at the speed of most conversations. Isolated or discrete speech recognizers require the user to pause after each word, and are currently being replaced by continuous speech engines.

Continuous speech recognition engines currently support two modes of speech recognition:

- *Dictation*, in which the user enters data by reading directly to the computer
- *Command and control*, in which the user initiates actions by speaking commands or asking questions

Dictation mode allows users to dictate memos, letters, and e-mail messages, as well as to enter data using a speech recognition dictation engine. The possibilities for what can be recognized are limited by the size of the recognizer's "*grammar*" or dictionary of words. Most recognizers that support dictation mode are speaker-dependent, meaning that accuracy varies based on the user's speaking patterns and accent. To ensure accurate recognition, the application must create or access a "*speaker profile*" that includes a detailed map of the user's speech patterns captured in the matching process during recognition.

Command and control mode offers developers the easiest implementation of a speech interface in an existing application. In command and control mode, the grammar (or list of recognized words) can be limited to the list of available commands (a much more finite scope than that of continuous dictation grammars, which must encompass nearly the entire dictionary). This mode provides better accuracy and performance, and reduces the processing overhead required by the application. The limited grammar also enables speaker-independent processing, eliminating the need for speaker profiles or "*training*" the recognizer.

The **command and control mode** is the one most adapted for speech commanding of robots.

4.2.4 Automatic Speech Recognition System and Strategy

From the several continuous speech ASR technologies available, based on personal computers, the *Microsoft Speech Engine* [4] was selected because it integrates very well with the operating systems we use for HMI, manufacturing cell control, and supervision (*Windows XP/NT/2000*). The *Microsoft Speech Application Programming Interface* (SAPI) was also selected, along with the *Microsoft's Speech SDK* (version 5.1), to develop the speech and text-to-speech software applications [4]. This API provides a nice collection of methods and data structures that integrate very well in the *.NET 2003* framework [5], providing an interesting developing platform that takes advantage of the computing power available from actual personal computers. Finally, the *Microsoft's SAPI 5.1* works with several ASR engines, which gives some freedom to developers to choose the technology and the speech engine.

Grammars define the way the ASR recognizes speech from the user. When a sequence included in the grammar is recognized, the engine originates an event that can be handled by the application to perform the planned actions. The SAPI provides the necessary methods and data structures to extract the relevant information from the generated event, so that proper identification and details are obtained.

There are three ways to define grammars: using XML files, using binary configuration files (CFG), or using the grammar builder methods and data structures. XML files are a good way to define grammars if a compiler and converter is available, as in the SDK 5.1. In the examples provided in this chapter, the grammar builder methods were used to programmatically construct and modify the grammar.

The strategy used here takes into consideration that there should be several robots in the network, running different applications. In that *scenario*, the user needs to identify the robot first, before sending the command. The following strategy is used,

- All commands start with the word "*Robot*"
- The second word identifies the robot by a number: one, two, etc
- The words that follow constitute the command and the parameters associated with a specific command

Consequently, the grammar used is composed of a "*TopLevelRule*" with a predetermined *initial state*, i.e., the ASR system looks for the pre-command word "*Robot*" as a precondition to any recognizable command string. The above mentioned sequence of words constitutes the *second level rules*, i.e, they are used by the *TopLevelRule* and aren't directly recognizable. A rule is defined for each planned action. As a result, the following represents the defined syntax of commands:

robot number command parameter_i

where "*robot*" is the pre-command word, *number* represents the robot number, *command* is the word representing the command to send to the robot, and *parameter_i* are *i* words representing the parameters associated with the *command*.

Another thing considered was safety. Each robot responds to "*hello*" commands, and when asked to "*initialize*" the robots require voice identification of username and password to give the user the proper access rights. Since the robots are connected to the calling PC using an RPC socket [2, 6-7] mechanism, the user must "*initialize*" the robot to start using its remote services, which means that an RPC connection is open, and must "*terminate*" the connection when no more actions are needed. A typical session would look like,

> **User**: Robot one hello.
> **Robot**: I am listening my friend.
> **User**: Robot one initialize.
> **Robot**: You need to identify to access my functions.
> **Robot**: Your username please?
> **User**: Robot one <*username*>.
> **Robot**: Correct.
> **Robot**: Your password please?
> **User**: Robot one <*password*>.
> **Robot**: Correct.
> **Robot**: Welcome again <*username*>. I am robot one. Long time no see.

Sequence of commands here. Robot is under user control.

> **User**: Robot one terminate.
> **Robot**: See you soon <*username*>.

In the following sections, two simple examples are given to demonstrate how this voice command mechanism is implemented, and how the robot controller software is designed to allow these features.

4.2.5 Pick-and-Place and Robotic Welding Examples

The following examples take advantage of developments done in the *Industrial Robotics Laboratory,* of the *Mechanical Engineering Department* of the *University of Coimbra* on robot remote access for command and supervision [2, 6-7]. Briefly, two industrial robots connected to an Ethernet network are used. The robot controllers (ABB S4CPlus) are equipped with RPC servers that enable user access from the network, offering several interesting services like variable access, IO access, program and file access and system status services [7]. The new versions of the ABB controller, named IRC5, are equipped with a TCP/IP sockets API [8], enabling users to program and setup TCP/IP sockets servers in the controller. For that reason, the ideas presented here can be easily transported to the new IRC5 controller with no major change.

If calls to those services are implemented in the client PC, it is fairly easy to develop new services. The examples presented here include the *ActiveX PCROBNET2003* [9] that implement the necessary methods and data structures (see Table 3.3) to access all the services available from the robot controller.

The basic idea is simple and not very different from the concept used when implementing any remote server. If the system designer can access robot program variables, then he can design his own services and offer them to the remote clients. A simple *SWITCH-CASE-DO* cycle, driven by a variable controlled from the calling client, would do the job:

```
switch (decision_1)
{
    case 0: call service_0; break;
    case 1: call service_1; break;
    case 2: call service_2; break;
    ...
    case n: call service_n; break;
}
```

4.2.6 Pick-and-Place Example

For example, consider a simple pick-and-place application. The robot, equipped with a two-finger pneumatic gripper, is able to pick a piece from one position (named "*origin*") and deliver it to other position (named "*final*"). Both positions are placed on top of a working table (Figure 4.4).

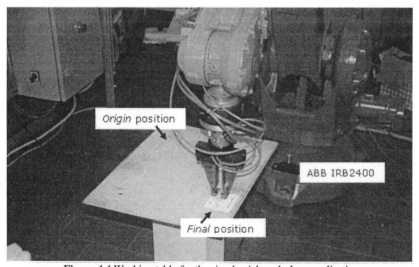

Figure 4.4 Working table for the simple pick-and-place application

The robot can be commanded to open/close the gripper, approach origin/final position (positions 100mm above origin/final position, respectively), move to origin/final position, and move to "*home*" (a safe position away from the table). This is a simple example, but sufficient to demonstrate the voice interface. Figure

4.5 shows a simplified version of the server software running on the robot controller.

To be able to send any of those commands using the human voice, the following grammar was implemented:

TopLevelRule = "*Robot*"	pre-command word
Rule 0 = "*one hello*"	check if robot is there
Rule 1 = "*one initialize*"	ask robot to initialize (*open client*)
Rule 2 = "*one master*"	rule defining username "*master*"
Rule 3 = "*one masterxyz*"	password of username "*master*"
Rule 4 = "*one open*"	open the gripper
Rule 5 = "*one close*"	close the gripper
Rule 6 = "*one motor on*"	put robot in run state
Rule 7 = "*one motor off*"	put robot in stand-by state
Rule 8 = "*one program run*"	start program
Rule 9 = "*one program stop*"	stop program
Rule 10 = "*one approach origin*"	call service 94
Rule 11 = "*one approach final*"	call service 93
Rule 12 = "*one origin*"	call service 91
Rule 13 = "*one final*"	call service 92
Rule 14 = "*one home*"	call service 90
Rule 15 = "*one terminate*"	release robot access (close client)

```
PROC main()
   TPErase; TPWrite "Example Server ...";
   p1:=CRobT(\Tool:=trj_tool\WObj:=trj_wobj);
   MoveJ p1,v100,fine,trj_tool\WObj:=trj_wobj;
   decision1:=123;
   WHILE TRUE DO
      TEST decision1
         CASE 90:
            MoveJ home,v200,fine,tool0; decision1:=123;
         CASE 91:
            MoveL final,v200,fine,tool0; decision1:=123;
         CASE 92:
            MoveL origin,v200,fine,tool0; decision1:=123;
         CASE 93:
            MoveJ Offs(final, 0,0,100),v200,fine,tool0; decision1:=123;
         CASE 94:
            MoveJ Offs(origin, 0,0,100),v200,fine,tool0; decision1:=123;
      ENDTEST
   ENDWHILE
ENDPROC
```

Figure 4.5 Simple pick-and-place server implemented in RAPID

The presented rules were introduced into a new grammar using the grammar builder included in the *Microsoft Speech API* (SAPI) [4]. The following (Figure 4.6) shows how that can be done, using the *Microsoft Visual Basic .NET2003* compiler.

```
TopRule = Grammar.Rules.Add("TopLevelRule",
SpeechLib.SpeechRuleAttributes.SRATopLevel Or
SpeechLib.SpeechRuleAttributes.SRADynamic, 1)

ListItemsRule = Grammar.Rules.Add("ListItemsRule",
SpeechLib.SpeechRuleAttributes.SRADynamic, 2)

AfterCmdState = TopRule.AddState
m_PreCommandString = "Robot"
TopRule.InitialState.AddWordTransition(AfterCmdState, m_PreCommandString, " ", , "",
0, 0)

AfterCmdState.AddRuleTransition(Nothing, ListItemsRule, "", 1, 1)
ListItemsRule.Clear()

ListItemsRule.InitialState.AddWordTransition(Nothing, "one hello", " ", , "one hello", 0, 0)
...
Grammar.Rules.Commit()
Grammar.CmdSetRuleState("TopLevelRule",SpeechLib.SpeechRuleState.SGDSActive)
RecoContext.State() = SpeechLib.SpeechRecoContextState.SRCS_Enabled
```

Figure 4.6 Adding grammar rules and compiling the grammar using SAPI in *Visual Basic .NET2003*

After committing and activating the grammar, the ASR listens for voice commands and generates speech recognition events when a programmed command is recognized. The corresponding event service routines execute the commanded strings. Figure 4.7 shows the shell of the application built in *Visual Basic .NET 2003* to implement the voice interface for this simple example. Two robots are listed in the interface. The robot executing the simple pick-and-place example is robot one (named *Rita*).

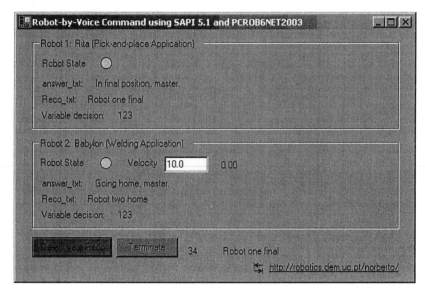

Figure 4.7 Shell of the voice interface application used to command the robot

With this interface activated, the following sequence of commands (admitting that the logging procedure was already executed) will take the robot from the *"home"* position, pick the work object at the origin position, deliver it to the final position, return to *"home"* and release the robot control.

User: Robot one approach origin.
Robot: Near origin, master.
User: Robot one open.
Robot: Tool open master.
User: Robot one origin.
Robot: In origin position master.
User: Robot one close.
Robot: Tool close master.
User: Robot one approach origin.
Robot: Near origin, master.
User: Robot one approach final.
Robot: Near final, master.
User: Robot one final.
Robot: In final position, master.
User: Robot one approach final.
Robot: Near final, master.
User: Robot one home.
Robot: In home position, master.
User: Robot one terminate.
Robot: See you soon master.

The speech event routine, running on the voice interface application, is called when any of the rules defined in the working grammar are recognized. For example, when the *"motor on"* rule is identified, the following routine is executed:

If ok_command_1 = 1 And (strText = "Robot one motor on") Then
    result1 = Pcrobnet2003.MotorON2(1)
    If result1 >= 0 Then
        Voice.Speak("Motor on, master.")
        ans_robot_1.Text() = "Motor ON, master."
    Else
        Voice.Speak("Error executing, master.")
        ans_robot_1.Text() = "Error executing, master."
    End If
End If

To give another example, when the move to "origin" rule is recognized, the following routine is executed:

If ok_command_1 = 1 And (strText = "Robot one origin") Then
    Dim valor As Integer
    valor = 92
    result1 = Pcrobnet2003.WriteNum2("decision1", valor, 1)
    If result1 >= 0 Then
        Voice.Speak("In origin position, master.")
        ans_robot_1.Text() = "In origin position, master."
    Else
        Voice.Speak("Error executing, master.")
        ans_robot_1.Text() = "Error executing, master."
    End If
End If

4.2.7 Robotic Welding Example

The welding example presented here extends slightly the functionality of the simple server presented in Figure 4.5, just by adding another service and the necessary routines to control the welding power source. The system used for this demonstration is composed of an industrial robot ABB IRB1400 equipped with the robot controller *ABB S4CPlus*, and a MIG/MAG welding power source (*ESAB LUA 315A*). The work-piece is placed on top of a welding table, and the robot must approach point 1 (named *"origin"*) and perform a linear weld from that point until point 2 (named *"final"*). The system is presented in Figure 4.8. The user is able to command the robot to

- Approach and reach the point origin (P1)
- Approach and reach the point final (P2)

- Move to "*home*" position
- Perform a linear weld from point P1 (origin) to point P2 (final)
- Adjust and read the value of the welding velocity

These actions are only demonstration actions selected to show further details about the voice interface to industrial robots. To implement the simple welding server, it is enough to add the following welding service to the simple server presented in Figure 4.5:

CASE 94:
```
weld_on;
MoveL final,v200,fine,tool0;
weld_off;
decision1:=123;
```

where the routine "*weld_on*" makes the necessary actions to initiate the welding arc [2], and the routine "*weld_off*" performs the post welding actions to finish the welding and terminate the welding arc [2].

The welding server is running in robot 2 (named *babylon*), and is addressed by that number from the voice interface application (Figure 4.9). To execute a linear weld from P1 to P2, at 10mm/s, the user must command the following actions (after logging to access the robot, and editing the velocity value in the voice interface application – Figure 4.9) using the human voice:

User: Robot two approach origin.
Robot: Near origin master.
User: Robot two origin.
Robot: In origin position master.
User: Robot two velocity.
Robot: Velocity changed master.
User: Robot two weld.
Robot: I am welding master.
User: Robot two approach final.
Robot: Near final master.

Figure 4.9 shows the voice interface when robot two is actually welding along with a user equipped with a handset microphone to send voice commands to the robot. The code associated with the welding command is,

```
If ok_command_2 = 1 And (strText = "Robot two weld") Then
    Dim valor As Integer
    valor = 95
    result1 = Pcrobnet2003.WriteNum2("decision1", valor, 2)
    If result1 >= 0 Then
        Voice.Speak("I am welding, master.")
        ans_robot_2.Text() = "I am welding, master."
```

```
        Else
            Voice.Speak("Error executing, master.")
            ans_robot_2.Text() = "Error executing, master."
        End If
    End If
```

The code above writes the value 95 to the variable *"decision1"*, which means that the service *"weld"* is executed (check Figure 4.5).

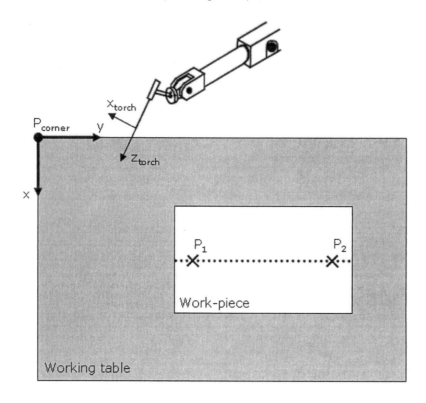

Figure 4.8 Simple welding application used for demonstration

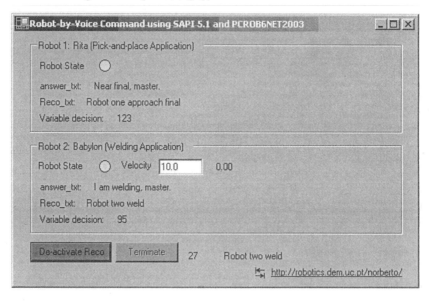

Figure 4.9 Shell of the voice interface application showing the welding operation, and a user (author of this book) commanding the robot using a headset microphone

4.2.8 Adjusting Process Variables

During the welding process, it may be necessary to adjust process variables such as the welding velocity, welding current, the welding points, and so on. This means that the voice interface must allow users to command numerical values that are difficult to recognize with high accuracy. Furthermore, it is not practical to define fixed rules for each possible number to recognize, which means that dictation capabilities must be active when the user wants to command numbers. To avoid noise effects, and consequently erroneous recognition, a set of rules were added to enable dictation only when necessary, having the rule strategy defined above always active. Consequently, the following rules were added for robot two (the one executing the welding example):

Rule V1 = "two variables"   enables access to variables
Rule V2 = "two variables out"   ends access to variables
Rule V3 = "two <*variable_name*>"   enables access to <*variable_name*>
Rule V4 = "two <*variable_name*> lock"   ends access to <*variable_name*>
Rule V5 = "two <*variable_name*> read"   reads from <*variable_name*>

Rule V6 = "two <*variable_name*> write"   writes to <*variable_name*>

Rules V1 and V2 are used to activate/deactivate the dictation capabilities, which will enable the easy recognition of numbers in decimal format (when the feature is activated, a white dot appears in the program shell – Figure 4.10). Rules V3 and V4 are used to access a specific variable. When activated, each number correctly recognized is added to the text box associated with the variable (a blinking LED appears in the program shell – Figure 4.10). Deactivating the access, the value is locked and can be written to the robot program variable under consideration. The rules V5 and V6 are used to read/write the actual value of the selected variable from/to the robot controller.

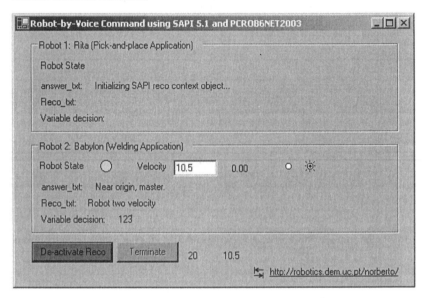

Figure 4.10 Accessing variables in the robot controller

As an example, to adjust the welding velocity the following code is executed after the corresponding rule is recognized:

If ok_command_2 = 1 And (strText = "Robot two velocity write") Then
Dim valor as Double
Dim velocity as Integer
valor = velocity.Text()
result1 = Pcrobnet2003.WriteSpeed("velocity", valor, 2)
If Result11 >= 0 Then
Voice.Speak("Welding velocity changed, master.")
ans_robot_2.Text() = "Welding velocity changed, master."
Else
Voice.Speak("Error executing, master.")

```
      ans_robot_2.Text() = "Error executing, master."
   End If
End If
```

Because the voice interface was designed to operate with several robots, two in the present case, the user may send commands to both robots using the same interface which is potentially interesting.

Using speech interfaces is a big improvement to HMI systems, for the following reasons:

- Speech is a natural interface, similar to the "*interface*" we share with other humans, that is robust enough to be used with demanding applications. It will change drastically how humans interface with machines
- Speech makes robot control and supervision possible from simple multi-robot interfaces. In the presented cases, common PC's were used, along with a quite normal noise-suppressing headset microphone
- Speech reduces the amount and complexity of different HMI interfaces, usually developed for each application. Since a PC platform is used, and they carry very good computing power, ASR systems become affordable and user-friendly

The experiments performed with this interface worked extremely well, even when high noise was involved (namely during welding applications), which indicates clearly that the technology is suitable to use with industrial applications where human-machine cooperation is necessary or where operator intervention is minimal.

4.2.9 Conclusion

In this section, a voice interface to command robotic manufacturing cells was designed and presented. The speech recognition interface strategy used was briefly introduced and explained. Two selected industrial representative examples were presented to demonstrate the potential interest of these human-machine interfaces for industrial applications.

Details about implementation were presented to enable the reader to immediately explore from the discussed concepts and examples. Because a personal computer platform is used, along with standard programming tools (*Microsoft Visual Studio .NET2003* and *Speech SDK 5.1*) and an ASR system freely available (SAPI 5.1), the whole implementation is affordable even for SME utilization.

The presented code and examples, along with the fairly interesting and reliable results, indicate clearly that the technology is suitable for industrial utilization.

4.3 VoiceRobCam: Speech Interface for Robotics

The example presented in this section extends the example in section 3.2, namely adding extra equipment and implementing a simple manufacturing cell-like system composed of a robot, a conveyor, and several sensors. It also includes a voice/speech interface developed to allow the user to command the system using his voice. The reader should consider the presented example as a demonstration of functionality because many of the options were taken with that objective in mind, rather than trying to find the most efficient solutions but instead the ones that suit better the demonstrating purpose.

The system (Figure 4.11) used in this example is composed of:

- An industrial robot ABB IRB140 [8] equipped with the new IRC5 robot controller
- An industrial conveyor, fully equipped with presence sensors, and actuated by an electric AC motor managed through a frequency inverter. To control the conveyor, an industrial PLC (*Siemens S7-200*) [12] is used
- A Webcam used to acquire images from the working place and identify the number and position of the available objects. The image processing software runs on a PC offering remote services through a TCP/IP sockets server

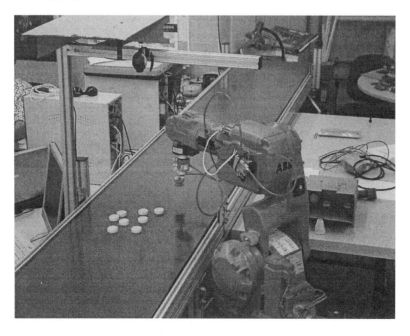

Figure 4.11 Manufacturing cell-like setup: picture and *Solidworks* model

In the following, a brief explanation of how the various subsystems work is provided. In the process, the relevant details about each subsystem and their respective construction are also given.

4.3.1 Robot Manipulator and Robot Controller

The ABB IRB140 (Figure 4.12) is an anthropomorphic industrial robot manipulator designed to be used with applications that require high precision and repeatability on a reduced working place. Examples of those types of applications are welding, assembly, deburring, handling, and packing.

ABB IRB 140 Basic Details

Year of release: 1999
Repeatability: +/- 0.03mm
Payload: 5kg
Reach: 810mm
Max. TCP Velocity: 2.5m/s
Max. TCP Acceleration: 20m/s2
Acceleration time 0-1m/s: 0.15 seconds

Figure 4.12 Details about the industrial robot manipulator ABB IRB140

This robot is equipped with the new IRC5 robot controller from *ABB Robotics* (Figure 4.13). This controller provides outstanding robot control capabilities, programming environment and features, along with advanced system and human machine interfaces.

IRC5 Basic Details

Year of release: 2005
Multitask system
Multiprocessor system
Powerful programming language: RAPID
FieldBus scanners: Can, DeviceNet, ProfiBus, Interbus
DeviNet Gateway: Allen-Bradley remote IO
Interfaces: Ethernet, COM ports
Protocols: TCP/IP, FTP, Sockets
Pendant: WinCE based teach-pendant
PLC-like capabilities for IO

Figure 4.13 Details about the industrial robot controller IRC5

The robot is programmed in this application to operate in the same way as explained in section 3.3.1, i.e., a TCP/IP socket server is available that offers services to the remote clients (see Table 3.3). This server is independent of the particular task designed for the robot, and allows only the remote user to send commands and influence the running task. In this case, the task is basically to pick objects from the conveyor and place them on a box. The robot receives complete commands specifying the position of the object to pick. Furthermore, since the relevant robot IO signals are connected to the PLC, the robot *status* and any IO action, like *"MOTOR ON/OFF"*, *"PROGRAM RUN/STOP"*, *"EMERGENCY"*, etc., are obtained through the PLC interface.

4.3.2 PLC Siemens S7-200 and Server

The PLC (Figure 4.14) plays a central role in this application, as it is common in a typical industrial manufacturing setup where the task of managing the cell is generally done by a PLC. In this example, to operate with the PLC, a server was developed to enable users to request PLC actions and to obtain information from the setup. To make the interface simple and efficient, the server accepts TCP/IP socket connections, offering the necessary services to the client's applications. The list of available services is presented in Table 4.1. The client application just needs to connect to the PLC server software application to be able to control the setup and obtain status and process information.

The server application (Figure 4.15) runs on a computer that is connected to the PLC through the RS232C serial port, and to the local area network (LAN) for client access.

Table 4.1 Services available from the PLC TCP/IP server

Service	Answer	Description
Init_Auto	<Init_Auto	Conveyor in Automatic Mode
Init_Manual	<Init_Auto	Conveyor in Manual Mode
Stop	<Stop>	Conveyor in STOP Mode
Read_Mode	Auto, Manual e Stop	Returns the conveyor operating mode
Manual_Forward	Manual_Forward	Conveyor starts in Manual Mode
Manual_Stop	Manual_Stop	Conveyor stops in Manual Mode
Force_Forward	<Force_Forward	Forces the conveyor to Start, although in Automatic Mode
IO	Bit stream*	Returns the status of all IO signals
Status	Bit stream**	Returns the status of all IO signals and the conveyor operating mode
Motor_On	<Motor_On>	Robot Motor ON
Motor_Off	<Motor_Off>	Robot Motor OFF
Prg_Run	<Prg_Run>	Robot Program RUN
Prg_Stop	<Prg_Stop>	Robot Program STOP

* The IO bit stream is formated in the following format:

BQ0.0:xxxxxxxxBQ1.0:xxxxxxxx BI0.0:xxxxxxxx:BI1.0:xxxxxxxx

where "*BQ0.0:*"/"*BI0.0:*" is string followed by 8 bits corresponding to the first block of digital outputs/inputs of the PLC, "*BQ1.0:*"/"*BI1.0:*" is a string followed by the 8 bits corresponding to the second block of digital outputs/inputs. For example, the following answer is obtained when BQ0.2, BQ1.0, BQ1.4, BQ1.6, BI0.1, BI1.0, BI1.1 and BI1.2 are activated:

BQ0.0:00100000BQ1.0:10001010BI0.0:01000000:BI1.0:11100000

Note: The bit assignment is as follows:

BQ0	0.0	0.1	0.2	0.3	0.4	0.5	0.6	0.7
	Conv_F	Conv_B	user	M_on	user	P_run	P_stop	M_off
BQ1	1.0	1.1	1.2	1.3	1.4	1.5	1.6	1.7
	user	user	user	user	user	user	user	User
BI0*	0.0	0.1	0.2	0.3	0.4	0.5	0.6	0.7
	Auto	Manual	M_on	M_off	P_run	P_stop	EMS	Busy
BI1	1.0	1.1	1.2	1.3	1.4	1.5	1.6	1.7
	Sensor1	Sensor2	Sensor3	User	user	user	user	user

*BI0 contains robot status information as listed.

** Similar to the above bit stream, but with the string "*Auto*", "*Manual*", or "*Stop*" added in the end of the stream in accordance with the state of the conveyor. For example, for the above mentioned IO state and with the conveyor in *Automatic Mode*, the answer to the *Status* call is,

BQ0.0:00100000BQ1.0:10001010BI0.0:01000000:BI1.0:11100000_Auto

Figure 4.14 Electrical panel showing the PLC, the frequency inverter and the electrical connections

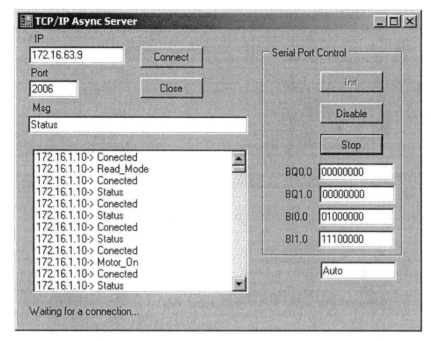

Figure 4.15 Shell of the PLC TCP/IP socket server

The PLC works as a server, as explained in Section 3.2.1.2, offering the IO services and actions necessary to control the system and obtain status information.

4.3.3 Webcam and Image Processing Software

This setup uses a simple USB *Webcam* to obtain images from the working area and compute the number of objects present and their respective positions. The camera is connected to a PC that runs the image processing software developed in *LabView* from *National Instruments* using the *IMAQ Vision toolbox*. The software works in the same way as explained in Section 3.3.2. Nevertheless, two more messages were added to the TCP/IP server, which return's the information necessary to calibrate the camera and to compute the object position in the robot's cartesian space (Table 4.2).

Table 4.2 Services from the Webcam TCP/IP server

Service	Description
camera get objects	Gets a frame from the *Webcam*
calibration pixels	Correlation between pixels and millimeters
cam to pos X_Y	Offset to add to the (x, y) position obtained from the image to compute the position of the object in the robot Cartesian space

The image processing software waits for a *"camera acquire objects"* message from the user client. When a message arrives, the server acquires a frame (image) from the camera and performs a binary operation, i.e., from a color image, or with several levels of gray, a back-and-white image is obtained with only two colors: black (0) or white (1). This type of procedure is necessary to identify the working objects in the scene and remove the unnecessary light effects.

The next task is to remove all the objects that are out of the working range. Those correspond to the parts of the conveyor belt, light effects, shadows, etc., and need to be removed before identifying the objects and computing their position.

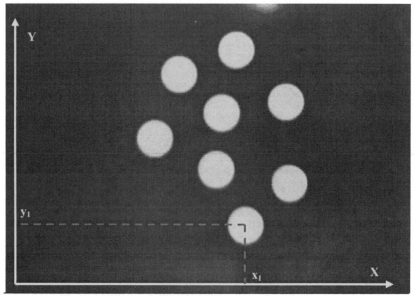

Figure 4.16 Frame obtained from the camera after being processed

Because the objects used with this application are small discs without holes (Figure 4.16), the image processing software uses a procedure to fill all the holes resulting from the binary operation. After that, a valid object should have a number of pixels between certain limits. This will allow users to identify unknown objects or objects that are overlapped. Only objects that pass this identification are considered, and for those the center of mass is computed: All other objects are ignored. From that the (x, y) position is computed and returned to the client application that issued the call.

4.3.4 User Client Application

It is now easy to understand the software architecture designed for this application (Figure 4.17): distributed and based on a client-server model. The user client application just need's to implement calls to the available services, track the answers, and monitor the robot and conveyor operations with the objective of controlling the setup.

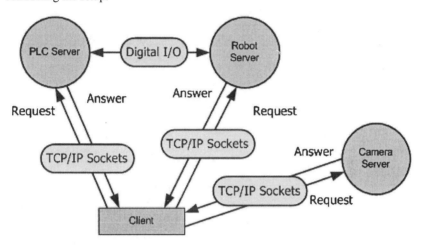

Figure 4.17 Basic distributed software architecture and connections between the different software modules

Figure 4.18 shows the shell of a PC client application developed using *C# .NET 2005* to access the above mentioned TCP/IP services from the various servers, and control the manufacturing cell-like system. With this application, the user can operate the setup in *"Manual Mode"*, i.e., issue all the actions independently, and at a time. The user can also have the conveyor in *"Automatic Mode"* and command the pick-and-place operation manually, i.e., require *"camera get objects"* to obtain the number of objects and their respective positions, selecting from the obtained list of objects the ones to pick.

Finally, the user can command the setup to work in fully *"Automatic Mode"*, i.e., to start the conveyor when objects are detected by sensor 1 (Figure 4.11), stop the conveyor when objects are detected by sensor 2, acquire an image of the working space and identify the number of objects and their positions, and then pick-and-place all of them and resume the conveyor operation.

For example, the "*Read IO*" and "*Motor ON*" actions are implemented with the following code:

Read IO	Send the message "IO" to the PLC TCP/IP socket server and process the returned answer

```
private void bt_ReadIO_Click(object sender, EventArgs e)
{
  int rec_num; string str_temp;
  m_socClient1 = new Socket(AddressFamily.InterNetwork, SocketType.Stream,
ProtocolType.Tcp);
  IPEndPoint remoteEP_PLC = new IPEndPoint(IPAddress.Parse("172.16.63.9"), 2006);
  m_socClient1.Connect(remoteEP_PLC);
  m_socClient1.Send(System.Text.Encoding.ASCII.GetBytes("IO<E>"));
  byte[] recData = new byte[256];
  rec_num = m_socClient1.Receive(recData);
  m_socClient1.Close();
  if (recData[6] == 48)
  {
    tapete = false;
    Tapete_ON.Checked = false;
  }
  else
  {
    conveyor = true; conveyor_ON.Checked = true;
  }
  if (recData[34] == 48)
  {
    sensor1 = false; sensor1.Checked = false;
  }
  else
  {
    sensor1 = true; sensor1.Checked = true;
  }
  if (recData[35] == 48)
  {
    sensor2 = false; sensor2.Checked = false;
  }
  else
  {
    sensor2 = true; sensor2.Checked = true;
  }
  if (recData[36] == 48)
  {
    sensor3 = false; sensor3.Checked = false;
  }
  else
  {
    sensor3 = true; sensor3.Checked = true;
  }
  str_temp = System.Text.Encoding.ASCII.GetString(recData, 0, rec_num);
}
```

Presenting the received information

Motor ON	Send the message that commands the robot "*Motor ON*" action

```
private void bt_Motor_ON_Click(object sender, EventArgs e)
{
  m_socClient1 = new Socket(AddressFamily.InterNetwork, SocketType.Stream,
ProtocolType.Tcp);
  IPEndPoint remoteEP_PLC = new IPEndPoint(IPAddress.Parse("172.16.63.9"), 2006);
  m_socClient1.Connect(remoteEP_PLC);
  m_socClient1.Send(System.Text.Encoding.ASCII.GetBytes("Motor_On<E>"));
  byte[] recData = new byte[256];

  m_socClient1.Receive(recData);
  m_socClient1.Close();
}
```

Figure 4.18 Shell of a client application developed in C# to control the setup (Sensor1 = "*Input Sensor*", Sensor2 = "*Cam Sensor*" and Sensor3 = "*Output Sensor*")

The client code is very simple and is composed of five main parts:
- Established socket client connection
- Send the command message
- Receive the answer
- Process and present the returned information
- Close the socket

When operating in fully "*Automatic Mode*" follows the sequence represented in Figure 4.19, which corresponds to the normal (or production-like) operation of the system. Considering a real production setup, it could be interesting to have more portable solutions. Consequently, a client application (Figure 4.20) was also developed to run on a *Pocket PC* (PPC). This application has the same basic functionality of the PC application (Figure 4.18).

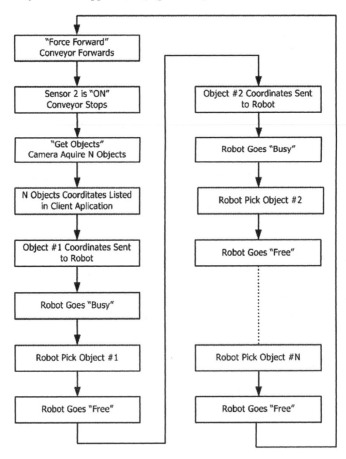

Figure 4.19 Sequence for the fully "*Automatic Mode*"

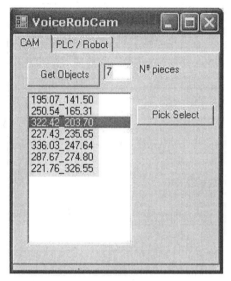

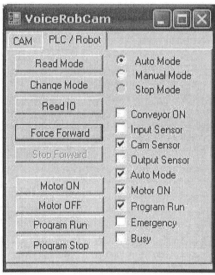

Figure 4.20 Aspects of the PPC client application developed in C# to control the manufacturing cell-like setup (Sensor1 = "*Input Sensor*", Sensor2 = "*Cam Sensor*" and Sensor3 = "*Output Sensor*")

4.3.5 Speech Interface

The current example is an interesting platform to demonstrate the potential of developing speech recognition systems for human-machine interfaces in industrial manufacturing systems. This statement is based on the following arguments:

- The system is constituted exclusively of industrial equipment, which makes it representative of a typical robotic manufacturing cell
- The software architecture developed to handle the system is distributed and based on a client-server model. This is a current trend in actual manufacturing plants
- The system uses industrial standards for network communications (Ethernet and TCP/IP)
- The system software was developed using commonly available software tools: *Microsoft Visual Studio .NET 2005*
- The concepts and technologies used in the system, for software, communications system organization, etc., are commonly accessible and most of them are currently defined as standards

As explained earlier, the system can be commanded manually, i.e., the various subsystems that compose the system can be directly commanded by the user. That perspective, or mode of operation, is explored in this section to introduce and demonstrate the enormous potential of current speech recognition (ASR) and text-to-speech (TTS) engines. In the presented implementation, the *Microsoft Speech API 5.1* (SAPI 5.1) is used to add speech recognition features (*speech commands*) to any of the above presented applications.

The strategy used to build the speech recognition grammar is simple and based on the concepts already presented in section 4.2. Since the system used here is composed of three different subsystems, a pre-command string per each piece of equipment is needed in the speech grammar. This allows the user to address each subsystem by its name,

> *m_PreCommandString1* = *"Robot"*
> *m_PreCommandString2* = *"Conveyor"*
> *m_PreCommandString3* = *"Camera"*

These three words (*"Robot"*, *"Conveyor"*, and *"Camera"*) are added to the speech recognition grammar as *TopLevelRules*, i.e., those words need to be identified to start the recognition of a command string. This means that the speech recognition grammar is built considering that the user commands have the following structure:

> *name_of_subsystem command parameters*

where "*name_of_subsystem*" is one of the *TopLevelRules*, i.e., one of the words that identify each of the subsystems, "*command*" is a string identifying the command, and "*parameters*" is a string containing the parameters associated with the specified command. Consequently, to have the system responding to speech commands, it is necessary to first identify the commands of interest and their associated parameters (Table 4.3).

Table 4.3 Commands associated with the speech command interface

TopRule	Robot	
Command	Parameters	Description
Hello	--	Checks if the speech recognition system is ready
Initialize	--	Initializes the interface and starts the login procedure, requesting *username* and *password*
Terminate	--	Terminates the speech interface.
<username>	--	Validates the "*username*"
<password>	--	Validates the "*password*"
Motor	On	Robot in Motors On state
	Off	Robot in Motors Off state
Program	Run	Starts loaded program from the beginning
	Stop	Stops loaded program
	Run from point	Starts loaded program from the actual program point
Program Option	Root	Selects program option "*root*": start menu
	<Number>	Selects program option defined by "*number*"
Pick	<Number>	Pick object defined by "*number*"
TopRule	Conveyor	
Command	Parameters	Description
Auto	--	Places conveyor in Automatic Mode
	Start	Forces the conveyor to start moving
Manual	--	Places conveyor in Manual Mode
	Start	Conveyor starts moving
	Stop	Conveyor stops moving
TopRule	Camera	
Command	Parameters	Description
Get Objects	--	Returns the number of objects in the scene and their respective positions
Calibration Pixels		Returns the pixel to millimeters ratio
Cam to Pos X_Y		Returns the offset that should be added to the computed positions to obtain the position in the robot Cartesian space

Therefore, adding the above presented rules to the speech recognition grammar (using an XML file or directly in the code), the ASR mechanism fires events when a rule is correctly identified. Consequently, the client application should just track the ASR generated events, discriminate the rule that was identified, and execute the associated actions. To perform those tasks, the ASR API provides functions that return the identified rule as a string. The application just needs to compare the string with the relevant possibilities, activating the associated actions when a match is obtained. Figure 4.21 shows some detail about the code associated with adding a speech commanding interface to the current application. Only the relevant parts of the code are listed, taking, as example, a few selected functions.

Speech Recognition Event Routine

```
...
   strText = Result.PhraseInfo.GetText(0, -1, True)
...
   If ok_command_1 = 0 And (strText = "Robot initialize") Then
      Voice.Speak("Your Username please?")
      ans_robot_1.Text() = "Your Username please?"
      ok_command_1 = -1
      asr_state.Text() = "Username."
   End If
...
   If ok_command_1 = -1 And (strText = "Robot master") Then
      Voice.Speak("Correct. And your password please?")
      ans_robot_1.Text() = "Correct. And your password please?"
      ok_command_1 = -2
      asr_state.Text() = "Password."
   End If
...
   If ok_command_1 = -2 And (strText = "Robot access level three") Then
      Voice.Speak("Correct. Welcome again master. Long time no see.")
      ans_robot_1.Text() = "Correct. Welcome again, master. Long time no see."
      ok_command_1 = 1
      If (result1 >= 0) Then
         robot1_on.Visible() = True
         asr_state.Text() = "Login OK."
      End If
   End If
...
   If ok_command_1 = 1 And (strText = "Robot terminate") Then
      Voice.Speak("See you soon, master.")
      ans_robot_1.Text() = "See you soon, master."
      s.Close()
      ok_command_1 = 0
      If (robot1_on.Visible = True) Then
         robot1_on.Visible = False
         asr_state.Text() = "Logout."
```

```
    End If
End If

If ok_command_1 = 1 And (strText = "Robot motor on") Then
    s = ConnectSocket(server_name, server_port)
    If s Is Nothing Then
        ans_robot.Text() = "Error connecting to robot, master"
        Voice.Speak("Error connecting to robot, master")
    Else
        Dim bytesSent As [Byte]() = Nothing
        bytesSent = ascii.GetBytes("motor_on")
        s.Send(bytesSent, bytesSent.Length, 0)
        'Voice.Speak("Motor on command received, master.")
        bytes = s.Receive(bytesReceived, bytesReceived.Length, 0)
        If Encoding.ASCII.GetString(bytesReceived, 0, bytes) = "0" Then
            Voice.Speak("Motor on, master.")
            ans_robot_1.Text() = "Motor on, master."
        Else
            Voice.Speak("Error executing, master.")
            ans_robot_1.Text() = "Error executing, master."
        End If
    End If
End If

If ok_command_1 = 1 And (strText = "Robot pick eight") Then
    s = ConnectSocket(server_name, server_port)
    If s Is Nothing Then
        ans_robot.Text() = "Error connecting to robot, master"
        Voice.Speak("Error connecting to robot, master")
    Else
        Dim bytesSent As [Byte]() =
        ascii.GetBytes("command_str 5000_" + object_cam(8))
        s.Send(bytesSent, bytesSent.Length, 0)
        bytes = s.Receive(bytesReceived, bytesReceived.Length, 0)
        ans_robot.Text() = Encoding.ASCII.GetString(bytesReceived, 0, bytes)
        's.Close()
        If Encoding.ASCII.GetString(bytesReceived, 0, bytes) = "0" Then
            Voice.Speak("Robot pick, master.")
            ans_robot_1.Text() = "Robot pick, master."
        Else
            Voice.Speak("Error executing, master.")
            ans_robot_1.Text() = "Error executing, master."
        End If
    End If
End If
```

```
If ok_command_1 = 1 And (strText = "Conveyor manual start") Then
    s = ConnectSocket(server_name_plc, server_port_plc)
    If s Is Nothing Then
        ans_robot.Text() = "Error connecting to conveyor, master"
        Voice.Speak("Error connecting to conveyor, master")
    Else
        Dim bytesSent As [Byte]() = ascii.GetBytes("Manual_Forward<E>")
        s.Send(bytesSent, bytesSent.Length, 0)
        bytes = s.Receive(bytesReceived, bytesReceived.Length, 0)
        pdata.Text() = Encoding.ASCII.GetString(bytesReceived, 0, bytes)
        Voice.Speak("Conveyor manual start, master.")
        ans_robot_1.Text() = "Conveyor manual start, master."
    End If
End If


If ok_command_1 = 1 And (strText = "Conveyor auto start") Then
    s = ConnectSocket(server_name_plc, server_port_plc)
    If s Is Nothing Then
        ans_robot.Text() = "Error connecting to conveyor, master"
        Voice.Speak("Error connecting to conveyor, master")
    Else
        Dim bytesSent As [Byte]() = ascii.GetBytes("Force_Forward<E>")
        s.Send(bytesSent, bytesSent.Length, 0)
        bytes = s.Receive(bytesReceived, bytesReceived.Length, 0)
        pdata.Text() = Encoding.ASCII.GetString(bytesReceived, 0, bytes)
        Voice.Speak("Conveyor automatic start, master.")
        ans_robot_1.Text() = "Conveyor automatic start, master."
    End If
End If
```

Figure 4.21 Detail about the code used in the ASR event routine

With this type of procedure, it is fairly simple add speech recognition features to the client applications described in this section. In general terms, the following is necessary (or desirable) to use speech commanding with industrial manufacturing systems:

- The system must be distributed in terms of software and based on a client-server model
- All the necessary subsystems must implement some type of mechanism for remote access from remote clients: RPC, TCP/IP sockets, etc
- A clear definition of the commanding strings must be available for easy implementation in different environments
- The speech recognition grammar developed for the application must reflect the above definitions. The routines associated with the recognition events must implement the service calls (using the defined commanding strings) and process the answers

- Some type of access mechanism must be implemented for security and safety reasons
- Critical commands should require some type of confirmation to avoid damaging persons and parts
- A careful selection of the headset used to implement the speech interface must be done, namely selecting devices with noise reduction electronics and with a press-to-speak switch

With these basic guidelines, speech recognition can be successfully added to industrial systems, resulting in a speech-enabled human-machine interface that could be a valuable improvement in terms of operator adaptation to the system. This would then improve operator productivity and efficiency, which would then impact the overall competitiveness of the company.

4.4 CAD Interfaces

Since the vast majority of companies use CAD software packages to design their products, it would be very interesting if the information from CAD files could be used to generate robot welding programs. That is, the CAD application could be the environment used for specifying how the welding robots should execute the welding operation on the specified parts.

Furthermore, because most welding engineers are familiar with CAD packages, this could be a nice way to proceed. An application presented elsewhere [2, 13-14] enables the user to work on the CAD file, defining both the welding path and the approach/escape paths between two consecutive welds, and organize them into the desired welding sequence. When the definition is complete, a small program, written in *Visual Basic*, extracts motion information from the CAD file and converts it to robot commands that can be immediately tested for detailed tuning. A set of tools is then available to speed up the necessary corrections, which can be made online with the robot moving. After a few simulations (with the robot performing all the programmed motions without welding) the program is ready for production. The whole process can be completed in just some minutes to a few hours, depending on the size and complexity of the component to be welded, representing a huge reduction of programming and set up time. Besides, most of the work is really easy offline programming.

These issues are further researched elsewhere [2, 13-14]. The objective here is to focus on the CAD interface and on adding more functionality to the human-machine interface of welding robots. Here the parameterization approach will be used. With this approach, the welding information, extracted from the CAD model, is used to parameterize a generic existing robot program, *i.e.*, the welding routines are implemented as general as possible enabling the accommodation of the planned welding tasks. In the case presented here, the information extracted from the CAD file, and adjusted using the presented software tools, is stored in a "*.wdf*" file and sent to the robot controller using the option "*Send to Robot*" of the software tool. The information is sent in the form of single column matrices serialized by the sequence that must be followed, *i.e.*, each line of any matrix contains the information corresponding to a certain welding point. As already mentioned, the robot controller is organized as a server, offering a collection of services to the remote computer. Therefore, the following are examples of services implemented in the welding server, running on a ABBIRB1400 industrial robot equipped with the S4C+ robot controller (the same robot used in Section 4.2).

Service 9100 (Move_CRobot): this service is used to move the robot in the Cartesian space with the specified TOOL frame, in accordance with the commanded offsets: x, y, z, rx, ry, and rz, where (x, y, z) are the Cartesian offsets and (rx, ry, rz) are the rotation offsets about the tool axis x, y and z, respectively.

Service 9401 (Welding): this service is used to execute the welding sequence commanded to the robot.

Service 9301 (Simulation): this service is used to execute the welding sequence without igniting the arc, *i.e.*, the welding power source is not activated.

Service 9101 (Move_JRobot): this service is used to move the robot in the joint space in accordance with absolute joint angles commanded from the remote computer.

Consequently, the main routine of the welding server may be implemented as a simple *SWITCH-CASE-DO* cycle, driven by a variable controlled from the remote computer (Figure 4.22).

Looking into the code in more detail, it's easy to find out how it works and how it can be explored, but also how new functions can be added to the system. Let's consider for example the *Move_CRobot* service (Figure 4.22) that corresponds to the value 9100 of the variable *decision1*. To move the robot in the cartesian space, the following must be commanded from the remote computer.

1. Enter the service routine: to do that, the user must write the value 9100 to the numeric variable *decision1*. The method from the *PCROBNET2003/2005* software component used to command that task is:

    pcrob.WriteNum("decision1", 9100, channel);

where *channel* identifies the RPC socket open between the robot controller and the remote computer.

```
PROC main()
    TPErase; TPWrite "Welding Server ...";
    reset_signals;
    p1:=CRobT(\Tool:=trj_tool\WObj:=trj_wobj);
    MoveJ p1,v100,fine,trj_tool\WObj:=trj_wobj;
    joints_now:=CJointT();
    decision1:=123; varmove:=0;
    WHILE TRUE DO
        TEST decision1
        CASE 9100:
            x:=0; y:=0; z:=0; rx:=0; ry:=0; rz:=0; move:=0;
            p1:=CRobT(\Tool:=trj_tool);
            WHILE (decision1=9100) DO
                IF (move <> 0) THEN
                    p1:=RelTool(p1,x,y,z\Rx:=rx\Ry:=ry\Rz:=rz);
                    x:=0; y:=0; z:=0; rx:=0; ry:=0; rz:=0; move:=0;
                ENDIF
                IF varmove <> 198 THEN
                    MoveJ p1,v100,fine,trj_tool\WObj:=trj_wobj;
                ELSE
                    MoveL p1,v100,fine,trj_tool\WObj:=trj_wobj;
                ENDIF
            ENDWHILE
            decision1:=123; varmove:=0;
        CASE 9101:
            joints_now:=CJointT();
            WHILE decision1=9101 DO
                MoveAbsJ joints_now,v100,fine,trj_tool\WObj:=trj_wobj;
            ENDWHILE
            decision1:=123;
        CASE 9401:
            weld;
            decision1:=123;
            p1:=CRobT(\Tool:=trj_tool);
            MoveJ RelTool(p1,0,0,-200),v100,fine,trj_tool\WObj:=trj_wobj;
        CASE 9301:
            weld_sim;
            decision1:=123;
        ENDTEST
    ENDWHILE
ENDPROC
```

Figure 4.22 Simple welding server running on the robot controller

2. Define the type of motion: the user must specify what type of motion to perform to achieve the target point, *i.e.*, linear motion or coordinated joint motion. This is specified writing to the variable *varmove* (198 for joint coordinated motion and any other value for linear motion). For example, the command

    pcrob.WriteNum("varmove", 198, channel);

specifies joint coordinated motion, using the open RPC socket identified by the parameter *channel*.

3. Command the Cartesian and rotational offsets: the user must write the offsets to the corresponding variables. After that, when the user signals that the offsets are available (writing a value different than zero to the variable *move*), the robot moves to the position/orientation obtained by adding those offsets to the actual position, and waits for another motion command. For example, the sequence of commands necessary to move the robot 20 mm in the positive X direction and 10 mm in the negative Z direction should be:

    pcrob.WriteNum("x", 20, channel);
    pcrob.WriteNum("y", -10, channel);
    pcrob.WriteNum("move", 1, channel);   ◄──────── **robot moves now!**

where again *channel* identifies the open RPC socket.

4. Leave the service: to leave this service the user must write any value different from 9100 to the variable *decision1*. For example, the following command writes the value -1 to the numeric variable *decision1* and makes the robot program quit the *Move_CRobot* service:

    pcrob.WriteNum("decision1", -1, channel);

Finally, let's consider the service *Welding* (Figure 4.22) that corresponds to the value 9401 of the variable *decision1*. The simplified version of the code is presented in Figure 4.23.

It is clear from the presented code (Figure 4.23) that the user should command the *Welding* service to execute, after sending the matrices defining the welding sequence. This service commands the robot to follow exactly the command sequence, moving the robot and igniting or stopping the welding arc whenever in the presence of a welding or approach/escape trajectory, respectively.

The example shows clearly that there are considerable gains in terms of flexibility and agility when using distributed client-server software architecture to assist industrial welding operations [2], namely taking advantage of the powerful programming tools developed for personal computers. It also shows that actual CAD packages can be used for robot programming tasks with great advantage, which extend the interest of already largely utilized software tools.

```
PROC weld()
    weldon:=0; i:=1;
    WHILE ((decision1=9401) AND (i<=numberpoints) AND (i>=1)) DO
        weldpoint:=i;
        wd_iref:=trj_voltage{i}; feed_iref:=trj_current{i};
        wd_href:=trj_voltage{i}; feed_href:=trj_current{i};
        wd_ref:=trj_voltage{i}; feed_ref:=trj_current{i};
        IF (trj_type{i}=0) THEN
            weld_on;
            weldon:=1;                                          Welding definition
        ENDIF
        ppos:=trj{i}; pvel:=trj_vel{i};
        pzone:=trj_prec{i}; ptype:=trj_mode{i};                Move the robot
        move_gen;
        IF (weldon=1) AND ((i+1>numberpoints) OR (trj_type{i+1}=1)) THEN
            weld_off;
            weldon:=0;
        ENDIF
        i:=i+1;
    ENDWHILE
    IF (weldon=1) THEN
        weld_off;
        weldon:=0;
    ENDIF
ENDPROC


PROC move_gen()
    IF ptype=0 THEN
        MoveL ppos,pvel,pzone,trj_tool\WObj:=trj_wobj;
    ENDIF
    IF ptype=1 THEN
        MoveJ ppos,pvel,pzone,trj_tool\WObj:=trj_wobj;
    ENDIF
    IF ptype=2 THEN
        TPWrite "[MOVE_GEN]: MoveC not implemented.";
    ENDIF
ENDPROC
```

Figure 4.23 Code for the *Welding* service

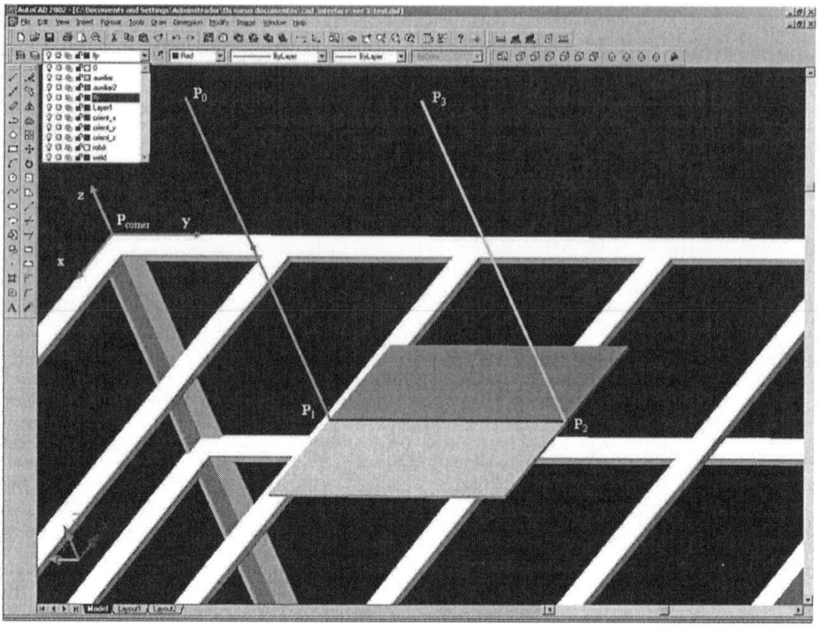

Figure 4.24 Definition of the simple welding example using *AUTOCAD*

To clarify further, let's consider the simple welding example already used in section 4.2.7. In that example, the robot is commanded to execute a linear welding on a work piece placed on a welding table. To demonstrate how this simple task is completely specified and programmed using a CAD package, the welding table and work piece were modeled in *AUTOCAD*. The same strategy used before is again utilized to specify points/orientations and trajectories, *i.e.*, they are all defined relative to a work object point/orientation (or reference system) named P_{corner}. In this way, when exporting points/orientations and trajectories to the robot, the only thing needed is a good calibration procedure of the robot TCP relative to P_{corner}, which can be done automatically using sensors (for example, laser position sensors) and special alignment routines, or manually using the robot joystick.

To execute the welding operation it is necessary to specify four points/orientations (P_0 to P_3) and the trajectories between them (Figure 4.24). The following procedures should be used:

1. P_0 should be defined as the approach point/orientation, *i.e.*, a point/orientation that could permit the robot to reach safely the work-piece from the "*home*" position. P_0 is consequently a non-welding point/orientation and the trajectory to P_0 should be free of obstacles (the user should guarantee adjusting P_0 accordingly). The precision to reach P_0 should be specified as low.

2. The trajectory from P_0 to P_1 should be defined as an approach linear trajectory, with point P_1 reached with the highest precision at low/medium velocity (let say 100mm/s, for example). As defined in [2], the weld layers in *AUTOCAD* are named for easy identification using a string that starts with the word "*WELD*". The next information is the type of trajectory, to distinguish between welding trajectories and approach/escape trajectories. After that should be specified the *welding current*, and then the *welding voltage*. Finally, the *welding speed* is specified. All these parameters are separated by spaces, constituting a definition string. Consequently, the label associated with that trajectory [2, 13-14] should be

WELD 1 0 0 0 100 0

for an approach/escape trajectory, done at 100mm/s with highest precision in the endpoint.

3. The trajectory from P_1 to P_2 should be defined as a welding trajectory with the required welding parameters. For example, the following label could be associated with this trajectory:

WELD 0 150.0 21.3 10 0

for a welding trajectory executed at 10mm/s, with highest precision in the end-point, associated with a welding current of 150.0 A and a welding voltage of 21.3 V.

4. The trajectory from P_2 to P_3 should be defined as an approach/escape trajectory done with low/medium velocity without any special precision in the endpoint. The following label could be associated with this trajectory:

WELD 1 0 0 0 100 50

to specify a trajectory done at 100mm/s, with low precision (50 mm sphere around the selected point).

This information is saved in the CAD file and can be extracted to a "*.wdf*" definition file, which is used for simulation and final tuning using the available tools [2, 13-14]. Finally, all of the information is sent to the robot using the already presented procedures, based on the routines developed for the robot controller and the "*write variable*" services (see Table 3.3) available from the *ActiveX* software [9] component used.

4.4.1 Speech Interface for Welding

Considering the linear weld case presented in Figure 4.24, a simple application was developed to command the welding procedure using a speech commanding interface. This is particularly relevant because the welding cells are usually very noisy and not attractive to operators, namely the younger ones. Consider that the trajectories were planned in *AUTOCAD* and transferred to the robot using the above mentioned applications. To operate the robot, the speech commands presented in Table 4.4 are necessary.

Table 4.4 Speech commands for the simple welding application

TopLevelRule	Robot	Number = Two
Command	**Parameters**	**Description**
Hello	--	Checks if the speech recognition system is ready
Initialize	--	Initializes the interface and starts the login procedure, requesting *username* and *password*.
Terminate	--	Terminates the speech interface.
<username>	--	Validates the *"username"*
<password>	--	Validates the *"password"*
Motor	On	Robot in Motors On state
	Off	Robot in Motors Off state
Program	Run	Starts loaded program from the beginning
	Stop	Stops loaded program
	Run from point	Starts loaded program from the actual program point
Approach	Origin	Approach *"Origin"* position
	Final	Approach *"Final"* position
Origin	---	Move to *"Origin"* position
Final	---	Move *"Final"* position
Weld	---	Perform a weld operation from *"Origin"* position to *"Final"* position

Note: The command message was defined in sections 4.2.4 and 4.2.7.

The application presented in Figure 4.10 implements a speech interface that recognizes those commands and executes the appropriate actions [2,13-14]. The user can command a welding operation just by saying:

User: Robot two approach origin.
Robot: Near origin, master.
User: Robot two origin.
Robot: In origin position, master.
User: Robot two weld.
Robot: I am welding, master.
User: Robot two approach final.
Robot: Near final, master.
User: Robot two home.
Robot: In final position, master.

That's easy, isn't it?
And it makes robotic welding a fun task. Like a computer game.

4.5 References

[1]   Observatory of European SMEs 2002, European Commission, 2002
[2]   Pires, JN, et al, "Welding Robots, Technology, System Issues and Applications", Springer, London, 2006
[3]   Pires, JN, "Semi-autonomous manufacturing systems: the role of the HMI software and of the manufacturing tracking software", Elsevier and IFAC Journal Mechatronics, to appear 2005.
[4]   Microsoft Speech Application Programming Interface (API) and SDK, Version 5.1, Microsoft Corporation, http://www.microsoft.com/speech
[5]   Microsoft Studio .NET 2003/2005, TechNet On-line Documentation, Microsoft Corporation, http://www.microsoft.com, 2003/2005.
[6]   Bloomer J., "Power Programming with RPC", O'Reilly & Associates, Inc., 1992.
[7]   RAP, Service Protocol Definition, ABB Flexible Automation, 1996 - 2004.
[8]   ABB IRC5 Documentation, ABB Flexible Automation, 2005.
[9]   Pires, JN, "PCROBNET2003, an ActiveX Control for ABB S4 Robots", Internal Report, Robotics and Control Laboratory, Mechanical Engineering Department, University of Coimbra, April 2004.
[10]   Pires, JN, "Complete Robotic Inspection Line using PC based Control, Supervision and Parameterization Software", Elsevier and IFAC Journal Robotics and Computer Integrated Manufacturing, Vol. 21, N°1, 2005
[11]   Pires, JN, "Handling production changes on-line: example using a robotic palletizing system for the automobile glass industry", Assembly Automation Journal, MCB University Press, Volume 24, Number 3, 2004.
[12]   Siemens, "S7-2000 Programmable Controller Programming Manual", Siemens Automation, Edition 08/2005, 2005.
[13]   Pires, JN, Godinho, Tiago, and Ferreira, Pedro, "CAD interface for Automatic Robot Welding Programming ", Volume 31, n°1, Industrial Robot, An International Journal, MCB University Press, 2004.
[14]   Pires, JN, and Loureiro, A, et al, "Welding Robots", IEEE Robotics and Automation Magazine, June, 2003

5

Industrial Manufacturing Systems

5.1 Introduction

Industrial small and medium (SME) manufacturing companies face complex and challenging market conditions that may impact their organization and economic strength. In fact, for a manufacturing SME to remain competitive in the global economy, it must cope with the following basic characteristics of the market:

- Global competition – actual companies compete on a global scale and with products from all over the world, i.e., coming from very different economic realities in terms of organization, labor, social protection and security, etc. Their competitors are global companies that address the markets with specific objectives and strategies, making the competition very unpredictable.
- Demand for more quality at lower prices – customers want the continuous improvement of quality at lower prices, i.e., customers tend to evaluate the quality of the product/service obtained for the money spent in buying it. This puts big pressure on companies since the market offers other options for the same product or service, and customers are used to making comparisons using the quality/price ratio.
- Very complex products – many of the modern high-technology products are very complex to manufacture since they often are composed of many mechanical parts, electronic components, software modules, etc. This poses new challenges to manufacturing systems.
- Very short life-cycles and time-to-market periods – competition and continuous innovation tends to reduce the life-cycle of products, forcing companies to evolve their line of products more often and with higher levels of agility.

This *scenario* poses very difficult challenges to manufacturing SMEs, namely on the quality of their manufacturing systems, in terms of flexibility and agility, and on their overall competitiveness. In fact, production plants based on human labor aren't competitive with equivalent companies located in low-salary countries. Consequently, these types of production plants tend to move their facilities to those countries or economical regions trying to take advantage of the low obligations to human labor, social security and protection, safety regulations, etc., and remain competitive in the global market. This logic has negative effects on western economies because important production sectors and jobs tend to move to low-salary countries. Consequently, the impact on the economic and social welfare is significant, working against our civilization model.

The only way to fight this trend is to focus on science and technology, developing manufacturing solutions that are flexible and agile, and that integrate efficiently with human operators. Flexibility is important to face the constant product change due to competition, fashion trends, quality requirements, and so on. But the time to market is also fundamental, which requires flexible systems that are easy to use and simple and fast to reconfigure, i.e., the modern world requires far more than flexibility and puts the focus on agility, which is a very interesting concept. Another important factor is the efficiency of the human-robot interfaces, which should allow humans and machines to operate as coworkers taking advantage of each other's abilities.

This chapter detail's a few industrial examples, with the objective of demonstrating how the concepts and ideas presented in this book can help to build manufacturing systems that are flexible, agile, and easy to use. All the systems presented were developed and built by the author of this book in cooperation with partner companies operating in Portugal.

5.2 Helping Wrapping Machines for the Paper Industry

In this section, a remote software environment developed to monitor and control robotic manufacturing cells is presented and discussed. It was used with an industrial system developed to wrap, label, and assist the storage of paper rolls coming from highly efficient paper machines. The system is also briefly introduced pointing out its main advantages. Special attention is given to the software architecture used to develop the remote services available from the system:

- Services for system monitoring
- Services for system maintenance
- Services for file and database handling
- Services for production monitoring
- Services for operator interface and system parameterization from the system control panel

The advantages of using distributed and object-oriented software approaches are also discussed, using some inside from the presented implementation. Finally, the utilization of electronic messaging services with industrial manufacturing systems is introduced and discussed.

5.2.1 Layout of the System

The system presented here was mainly designed to be used at the end of a paper machine to help with the wrapping and labeling operations of the paper rolls. Briefly, paper is produced in cylindrical rolls of several dimensions (with diameters ranging from 800mm up to 1600mm, and lengths ranging from a few centimeters to 2-3 meters) and weights. Figure 5.1 represents a diagram of the system showing its basic stations, i.e., places where robots are used to perform the required operations.

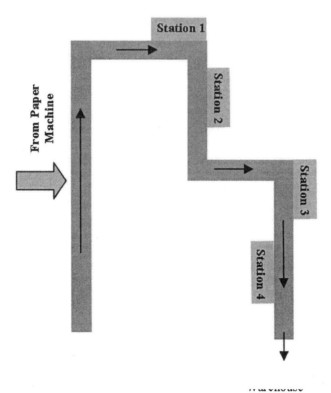

Figure 5.1 Basic organization of the of the robotic wrapping and labeling system

Paper rolls coming from the paper machine are labeled by a human operator using barcode sticks. The assigned code constitutes a unique identification of each roll. In the first station, the paper rolls are measured and weighted and that information is automatically inserted into the factory production database for further use, namely on the subsequent stations to pre-position the subsystems used in each station and to adapt the behavior of the local software. The system is controlled using industrial PLCs, which are accessible through *Profibus* by the PC that run's the human-machine interface software. The fieldbus network connecting the various system resources is also *Profibus*.

5.2.1.1 Station One – Dimensions and Weight
In this station, each roll is measured and weighted automatically and autonomously. The obtained values are introduced into the production database using the ID number in the barcode (barcode readers are used here). The rolls are serialized starting from this point and consequently there is no need to keep track of the rolls in the rest of the process, i.e., after this station there is no way to remove the rolls manually. The barcode numbers will be checked again at the end of the wrapping process when the rolls enter the automatic warehouse.

5.2.1.2 Station Two – Roll Wrapping and Inner Header
Rolls are wrapped using a wrapping machine assisted by two industrial robots ABB IRB6400 (equipped with the S4C+ robot controller) [1]. The robots are commanded to pick two headers, one per robot, of the appropriate dimensions (there are six piles of different headers available) and hold them against the two bases of the roll (Figure 5.2). The dimension of the header to pick is a parameter of the pick command, which is sent to each robot through *Profibus*. Consequently, a client-server software architecture is used, having the robots operating as servers. Synchronization and messaging (including error handling) with the station PLC, which also handles the wrapping machine, is done by *Profibus* using a simple IO protocol. The system is able to wrap rolls in cycles of less than 20 seconds.

a)

b)

Figure 5.2 Operation in station two: a) holding the headers, b) picking a header

5.2.1.3 Station Three – External Header

External headers are applied on the rolls to finish the roll wrapping process and hold the wrapping paper. Operation is assisted using one industrial robot (ABB IRB6400 equipped with the S4C+ robot controller) [1]. The robot is commanded to pick two headers (gripper holds two headers) and put them, properly centered in accordance with its diameter, on the plates of a heated press. The headers are made from a type of paper that has glue impregnated in its structure. The heat makes the glue emerge at the surface of the header, enabling the press to glue them to the rolls just by applying pressure (Figure 5.3). Due to the cycle time requirements (less than 20 seconds per roll), the command sent to the robot to pick a pair of headers includes the diameter of the actual roll (like in the previous station) but also the actual position of the press plates (to speed up the wrapping process, the press is independently commanded to pre-position its plates as a function of roll length). Since the press is hydraulic, the position of its plates is confirmed by the robot just before entering the press workspace to place the headers. This presents robot collisions with press plates, which would eventually destroy the robot and gripper.

5.2.1.4 Station Four – Labeling

In this station (Figure 5.4), two labels are applied to the wrapped rolls (one on the top and the other on the right side of the roll) with the information about the roll printed in the label (dimensions, weight, customer, production date, etc.). Each label also has a barcode that will be used by the automatic warehouse to process the roll. Labels are printed by an office laser printer, and outputted to a small ramp. The robot picks the labels when commanded to do it, waits for the *"glue labels"* command, puts glue on the surface of the labels (using the gluing machine), waits for the roll in position, and finally places the labels on the roll. After each basic operation, the execution status is checked and the next operation is commanded only if the previous one finished successfully. If an error occurs, the current process is aborted and the error is issued back to the commanding machine (in this case a PLC).

a)

b)

Figure 5.3 Operation in station three: a) picking a pair of headers, b) placing the headers on the surface of the heated plates of the hydraulic press

This same procedure is used in any of the other stations. All commands are acknowledged when they finish, i.e., a message specifying that the command executed correctly is sent back to the commanding machine. Communication runs over *Profibus* using a simple IO protocol.

Another version of this labeling station was built for another paper machine (see section 3.6), at the same company, that uses an *Ethernet* network and a PC to interface with the production database. The PC is also used to command the station, using *remote procedure calls* (RPC) sent to the robot controller. It is important to discuss here the basic differences between the two systems.

Considering the brief description made above and in section 3.6, and considering that robots used in industrial applications are commanded to execute very precise tasks, it is clear that in both cases there's the need for a collection of services properly designed to execute those tasks. Both systems implement a collection of services designed to execute every task available from the system. The services are implemented as generally as possible and require parameters to be properly requested by the remote client. A simple "*switch-case-do*" loop, driven by the word or number that defines the command, can be used to implement the server.

The difference resides in the way those services are requested. In the example presented in section 3.6, the services are requested using RPC calls, and in the example presented in this section the services are requested using a simple IO protocol (see section 3.2.1). Furthermore, the version presented in section 3.6 includes an intermediate server used to connect the factory production software and the robot controller (Figure 5.5). This server listens for TCP/IP calls and simply translates the calls to robot commands, collecting the answers and sending them back to the calling machine (the production software computer).

a)

b)

Figure 5.4 Labeling system: a) tool and gluing machine, b) Robot placing label

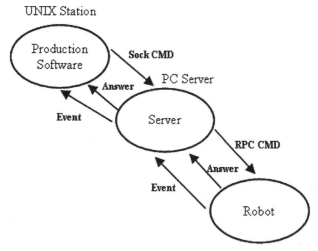

Figure 5.5 Connection between robots and factory production software: using TCP/IP sockets and SUN RPC 4.0 compatible RPCs

5.2.2 EmailWare

In every station presented, any error is logged and sent to the commanding computer as part of the answer: error codes are used to identify each type of error. Consequently, on an error situation the calling machine can decide what to do based on the received error code, for example, repeat the command.

Furthermore, every system has a checklist of basic conditions it needs to operate. For example, the labeling system needs to verify the following conditions to enter the *ready mode*:

- Air pressure at the appropriate working level
- Printing machine at the ready mode
- Glue machine at the ready mode

If a system is experiencing some type difficulty and one of the above conditions is not met, then the system enters the *"error mode"* and rejects all the incoming commands until the problem is solved.

At this point several things can be done. Let's discuss it a little bit more with an example. Suppose that there was a vacuum failure in the gripper, caused by air pressure failure (*venturi* devices are used to generate vacuum for suction cups). The system is then unable to pick and hold labels. If the problem appears during task execution, then an event may be fired (if an event firing mechanism is available) and an error code is issued back when the command finishes (0 – success, < 0 means an error identified by the error code). The simple way to proceed and to warn operators is to act on local warning devices (a bell, a flashing light, etc), on flashing warnings on system panels, etc.

This *scenario* was the motivation to develop the *EmailWare* application, which was then extended to enable a more general task of supervising and monitoring the complete system. With those ideas in mind, a server was built to monitor an installation of robots (networked robots using TCP/IP over *Ethernet* or a serial channel) inside a factory or in a research environment. The server uses the already mentioned *ActiveX* component (*PCROBNET2003/2005*) and is capable of checking the robots available on the nework for selected interesting information, logging all events, and warning the user immediately when a selected event actually occurs. Operators are not always near the system control computer, but can be reached by beeper, mobile phone, or e-mail. In fact, they can be in an office doing some desktop job, somewhere in the plant or at home after hours. A manufacturing system should be able to reach them to send urgent information. The same situation happens with developers. They need to recollect information about their systems and sometimes, on debugging situations, they need information when certain conditions are met.

One good solution is to use short e-mail messages sent to selected accounts with brief information about events. Those accounts could be regular e-mail accounts, SMS services, beepers, etc. The application should also accept e-mail messages, coming from authorized users requesting more details about any subject (see Tables 5.1 and 5.2).

Using this application, the user may define for each robot in the installation the type of events he wants to receive. The user can also request the system to send complete reports daily, weekly, or monthly. When one of the selected events actually occurs, the application sends a short e-mail to the defined e-mail accounts. The user also selects the accounts that can receive reports, log files, or long e-mails (long e-mail should not be sent to SMS accounts or beepers).

Table 5.1 Type of events

Type of event	Parameter 1	Parameter 2	Parameter 3	Parameter 4
IO_DIGITAL	name	T0 / T1		
IO_ANALOG	name	H / L	Value	
VAR_NUM	name	H / L / C	Value	
VAR_BOOL	name	T0 / T1		
STATE_SYS	TA / TM			
STATE_PRG	TR / TS			
ERROR				
LOGS_D	type	type	Type	...
LOGS_W				...
LOGS_M	type	type	Type	...

where the symbols have the following meaning:

IO_DIGITAL – digital IO events.
IO_ANALOG – analog IO events.
VAR_NUM – events related with RAPID <num> variables.
VAR_BOOL – events related with RAPID <bool> variables.
STATE_SYS – system state events.
STATE_PRG – program state events.
ERROR – error events (any type of error).
LOGS_D – send logs daily.
LOGS_W – send logs weekly.
LOGS_M – sends logs monthly.
name – name of variable or signal (string).
T0 – transition to zero.
T1 – transition to 1.
H – Higher than value.
L – Lower than value.
C – When variable changes.
TA – transition to auto mode.
TM – transition to manual mode.
TR – transition to program running.
TS – transition to program stop.
type – type of log.

Table 5.2 Type of commands

Command	Parameter 1	Parameter 2	Parameter 3
LOGS	type	type	...
SYSTEM			
PROGRAM			
IO_DIGITAL	all / signal	signal	
IO_ANALOG	all / signal	signal	
IO_ALL			
VAR_NUM	name		
VAR_BOOL	name		
STOP_PRG	password		
START_PRG	password	AP / FB	
UNLOAD	password	name	
LOAD	password	name	
MOTOR_ON	password		
MOTOR_OFF	password		
X_CMD	password	par_1	...

where the symbols have the following meaning:
LOGS – send log files.
SYSTEM – send system state information.
PROGRAM – send program state information.
IO_DIGITAL – send information about digital IO as specified.
IO_ANALOG - send information about analog IO as specified.
IO_ALL – send information about all IO.
STOP_PRG – stops current program.
START_PRG – starts current program.
UNLOAD – unload module specified (name).

LOAD – load module specified (name).
MOTOR_ON – motors ON state.
MOTOR_OFF – motors OFF state.
X_CMD – any command implemented in RAPID.
all – all signals of this type.
password – password to execute this command (if password fails, then user is removed from list of allowed users and an e-mail to administrator is issued).

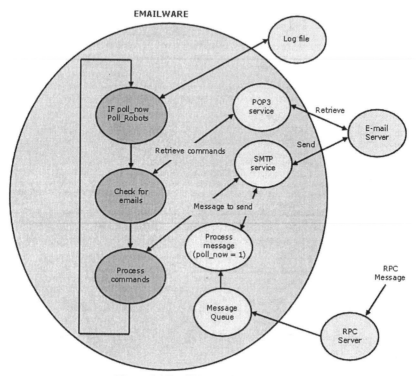

Figure 5.6 EmailWare: selecting a robot

Another important feature is the possibility to send e-mail commands to the application asking for more details on several aspects (see Table 5.2 for the types of commands that can be issued). The user can issue commands to any robot in the installation. The application checks if the sender is allowed and then processes the command. Those commands are e-mail messages sent to *emailware@company* with subject "*command*" and with the following syntax:

 # robot_dns_name command parameters

where "*robot_dns_name*" is the registered DNS name of the robot and "*command*" is a command, using the required "*parameters*" from Table 5.2. The e-mail message can hold any number of commands (one per line starting with character '#') addressed to several robots.

The application cycle polls all robots for any change (it does not keep open clients, just opens a client connection, makes a survey, and closes the connection), fires e-mails if there is any change, and then processes commands (Figure 5.6). Since there is an RPC server working in parallel receiving asynchronous messages from any robot, any urgent event is immediately attended and information is issued to the user (the information is sent once when it happens, i.e., when the event is fired from the robot, and a second time when the polling process detects the change). The polling frequency of the robots can be adjusted to avoid overloading the system, ranging from 1/10 Hz (higher frequency) to 1/60 Hz (lowest frequency).

5.2.2.1 EmailWare Application Example

To show the potential of this tool, lets give a simple example. Suppose that at some industrial installation there is a robot (named "*babylon5*") doing arc-welding operations. Suppose also that the welding software keeps information on the number of pieces that have been welded (*num_pieces*), on the amount of time in operation (*opr_time*), and on the idle time (*idle_time*). There is also information on how many errors were encountered during operation (*num_error*); it is considered here that the system can handle and maybe automatically recover from certain operational errors (consequently, for each error the *num_error* variable is incremented and an operational message is issued like: *bad* or *no piece in place*, *no gas*, *no air pressure*, etc), which is normally the case. There are also some IO inputs and outputs like: gas information (digital input, *gas_on*), air pressure information (digital input, *air_on*), wire information (digital input, *wire_on*), etc. Finally, suppose that the user wants to have daily reports about the system, including the state of some of variables.

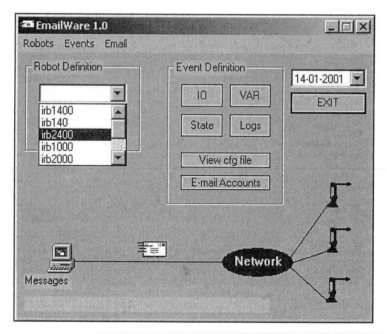

Figure 5.7 EmailWare: selecting a robot

To configure *EmailWare* for the welding application, the user starts by selecting the robot from the available robots (Figure 5.7). After that, the user selects the IO signals, the variables, and the type of system states of interest.

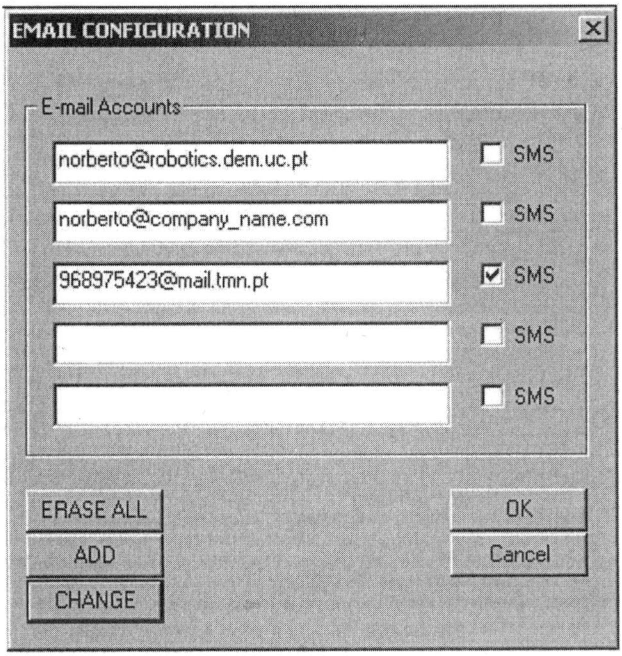

Figure 5.8 EmailWare: dialog to define e-mail accounts

Then the user e-mail accounts (Figure 5.8) must be defined (up to five accounts) and the ones that can receive long e-mails (the user should identify at least one normal e-mail account and one SMS account) must be specified. All the configurations are stored in a configuration file (*rob_conf.cfg*) that can be accessed using any text editor (*Notepad*, *Wordpad*, *Word*, etc). For the above-mentioned example, the file could look like the one in Figure 5.9.

As mentioned already, the application was tested on the industrial installation, presented in this section which uses four robots, but the interested reader can make his own test using our laboratory robots. Just visit the *EmailWare* web site located at *http://robotics.dem.uc.pt/emailware/* and sign up to receive warnings about the operation of one of our robots. Interested readers can also send commands to it. The site is a demonstration site, so only a few features are demonstrated and users cannot customize them. Finally, a demo version that is fully operational for one robot only (robot serial number is needed) may also be requested.

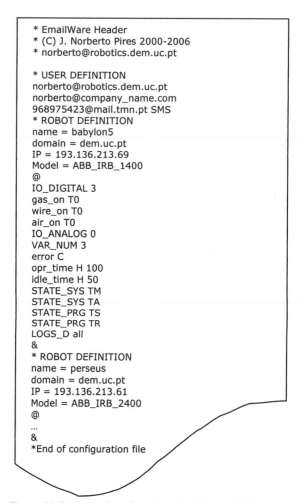

```
* EmailWare Header
* (C) J. Norberto Pires 2000-2006
* norberto@robotics.dem.uc.pt

* USER DEFINITION
norberto@robotics.dem.uc.pt
norberto@company_name.com
968975423@mail.tmn.pt SMS
* ROBOT DEFINITION
name = babylon5
domain = dem.uc.pt
IP = 193.136.213.69
Model = ABB_IRB_1400
@
IO_DIGITAL 3
gas_on T0
wire_on T0
air_on T0
IO_ANALOG 0
VAR_NUM 3
error C
opr_time H 100
idle_time H 50
STATE_SYS TM
STATE_SYS TA
STATE_PRG TS
STATE_PRG TR
LOGS_D all
&
* ROBOT DEFINITION
name = perseus
domain = dem.uc.pt
IP = 193.136.213.61
Model = ABB_IRB_2400
@
...
&
*End of configuration file
```

Figure 5.9 Example of configuration file (*rob_conf.cfg*)

Consequently, any of the specified users receive messages (by e-mail or SMS) about the programmed events that can look like:

Babylon 5: Ei guys, I'm stopped, no air-pressure or air-pressure too low.
Babylon 5: Ei guys, I'm stopped, no wire.
Babylon 5: Ei guys, wire is running out.
Babylon 5: OK, air-pressure is on again.

5.2.3 Conclusions and Discussion

The system presented in this section is commanded remotely from the PLC used to manage the operation of the cell. The system also uses a PC to interface with the operator, and updates and retrieves information from the factory production software. The system was designed to operate almost autonomously, i.e., with minor operator intervention limited to error and maintenance situations. Consequently, a client-server software architecture was used, with the robots working as servers allowing remote clients to explore and operate the system. This proved to be a nice solution capable of providing a good performance and high levels of flexibility, because the system's basic operation is defined by the operating software. Adding new functions or changing the operation is an easy task and in fact was done several times to adjust to new requirements.

Finally, a simple e-manufacturing solution was introduced in this section. It enables operators to receive operation events when they occur, allowing a more efficient supervision of the system, reducing down time due to errors or unavailability of certain operating conditions. This idea of having automation equipment sending messages to users with relevant information about its current status, and enabling users to request more details and sending a few commands, also by e-mail, can be extended to other areas: monitoring warehouse systems that could inform users about critical points, smart houses informing users about current situations and enabling some remote commands, remote maintenance, and so on.

5.3 Complete Robotic Inspection Line for the Ceramic Industry

Non-flat ceramic products, like toilets and bidets, are fully inspected at the end of the production process to search for structural, surface, and functional defects. Ceramic pieces are transported to the inspection lines assembled in pallets, carried by electro-mechanical fork-lifters or *automatic guided vehicles* (AGV). Pallets need to be disassembled, feeding the inspection lines where human operators execute the inspection tasks. Also, the pieces that pass inspection need to be palletized again in the final pallets used for product distribution. Those de-palletizing and palletizing operations are physically demanding so they are good candidates for robots.

This section is a case study on the development of a collection of prototype manufacturing cells, designed to perform automatic palletizing and de-palletizing operations of non-flat ceramic pieces such as toilets and bidets. The factories of these types of products show an impressive mixture of human and automatic labor, meaning that special attention must be taken with regard to human machine interfaces (HMI), safety, mode of operation, etc.

Non-flat ceramic products are commonly used in our homes and are mainly associated with personal care tasks. The industrial production of these ceramic products poses several problems to industrial automation, especially if robots are to be used. Basically, these problems arise from the characteristics of the ceramic pieces: non-flat objects with high reflective surfaces, very difficult to grasp and handle due to the external configuration, heavy and fragile, extensive surface sensitive to damage, high demand for quality on surface smoothness, etc. Also, the production setups for these types of products require high quality and low cycle times, since this is a large scale industry that will remain competitive only if production rates are kept high. Another restriction is that this industry changes products frequently, due to fashion tendencies in home decoration, etc. Also, there is the mixture of automatic and human labor production, which is a difficult problem since HMI are very demanding and a key issue in modern industrial automation systems.

It was proposed by the partner company to build several de-palletizing and palletizing solutions, with a simple graphic operator interface, to install in their final inspection lines. In those lines human operators inspect all pieces by hand to find functional and surface defects (computer vision solutions for inspection). The challenge was to build highly efficient systems, capable of handling more pieces a day than its human counterparts, that could be easy to set up and start up at the beginning of the day. So, there is a robotic challenge and a software challenge, namely, in designing human-machine interfaces for operators.

The system presented here (Figure 5.10) was designed to take advantage of computers and available tools to parameterize and monitor an industrial robotic cell, i.e., to make human-machine interface. In the process of describing and discussing the system a few available, a few technical details are highlighted. This is also important due to the fact that all the software was built from the scratch [2], without using any of the available commercial software packages (Section 3.2).

5.3.1 Motivation and Goals

The problem addressed in this example is the construction of a complete system to assist humans in the task of inspecting non-flat ceramic pieces. Those pieces (bidets and toilets, mainly) reach the inspecting site directly from the high temperature oven, organized in pallets (input-pallets), using fork-lifters. A few operators placed along two inspecting lines (15 meters long each), inspect all the pieces by hand, searching for pieces with functional and surface defects, removing from the inspection lines the pieces rejected [3, 4]. Consequently, in this system there is the need to de-palletize the input-pallets, feeding continuously the two inspection lines. The system must also pick the accepted pieces from the end of the inspection line, palletizing them again into the pallets (output-pallets) used for product distribution (Figure 5.10).

The system should work also as autonomously as possible, requiring only minor parameterization at the beginning of the work day or production cycle. The system should be able to work with input-pallets composed of four levels of ceramic pieces, eight pieces per level placed in a special order to keep pallet equilibrium, and with levels separated with pieces of hard paper. It should also be able to work with output-pallets up to five levels of ceramic pieces, eight pieces per level placed in the same order as in the input-pallets, with levels also separated by hard paper. The rule used to arrange the pieces in the pallet is to place them alternatively one up – one down, starting from the ground level, then swap to one down – one up in the next level (Figure 5.11), and keep the procedure in the proceeding levels.

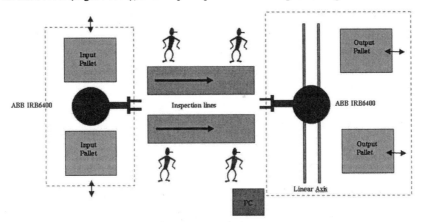

Figure 5.10 Components of the system

Actually, input-pallets are assembled manually by operators at the end of the high temperature oven. This means that the robotic system must be tolerant with possible medium-large palletizing errors, coming from misplaced pieces both in position and orientation, and also showing significant variations from level to level. Another important factor is that pallets are fed into the system by human operators using electro-mechanic fork-lifters, which also introduces some variation in the pallets. Sometime in the future, AGVs will be use to fulfill the task, reducing considerably the variations introduced and increasing the efficiency of the system.

a)

b)

c)

Figure 5.11 Pallets and view of the system: a) input pallets and de-palletizing robot; b) aspect of the de-palletizing gripper; c) view of the complete system

The main objectives for this system are summarized as follows:

- Build a complete robotic system capable of performing de-palletizing and palletizing operations to assist inspection lines
- The system must perform each of these operations in less than 12 seconds per piece
- The system should cope with high palletizing errors on the input-pallets, since they are assembled by human operators which permits to anticipate small-medium placement errors (up to 5cm in position and up to 5° around the vertical axis)
- The system should cope with deviations on the dimensions of the pieces of up to ±1 cm in each direction. Ceramic pieces grow inside the high temperature oven, making these deviations expected due to temperature deficiencies, variation of time inside the oven, variations in the ceramic mixture, etc. These deviations are not necessarily errors, but instead a characteristic of this type of production
- The system must work with pallets, both input and output, with variable numbers of pieces, ranging from any number of pieces, in the case of the input pallets, to 8, 16, 24 or 40 pieces, in the case of the output pallets

- The system should maintain information about its surroundings, so as to warn about inconsistencies between what is ordered and what is available
- The system must be parameterized easily, using a graphical interface implemented with a touch-screen. A few commercial software packages are available in the market. Nevertheless, our option was to build our own solution since the human-machine interface plays a crucial role in the performance of the system, including operator acceptance. It is therefore very important to have full control over the developed software
- The system must be optimized for each model of ceramic pieces. This means that there should be the option of introducing new models using a teach strategy

Considering these above mentioned objectives, the following challenges were identified:

- To build a human-machine interface, easy to use and capable of handling production needs. System warnings and errors must be issued to the operator's attention in an efficient way. All operations and messages must be logged for future analysis;
- To build a system capable of meeting the planned requirements;
- To explore the capabilities of the current personal computers, operating systems, and related tools on a very demanding industrial environment;

Taking the above objectives and challenges, and considering the fact that this is an industrial project, meaning it is supposed to work 24 hours a day without problems, it was decided to distribute the software to all the components of the system. A client-server architecture [2-8], based on remote procedure calls (RPC) [9], was adopted, with the PC as the client of the rest of the components of the system, including the robot controllers, and also as the interface to the operator.

5.3.2 Approach and Results

The objectives and requirements of this project necessitated a robotic cell that could handle the ceramic pieces under consideration. Proper grippers and layouts were designed and built. It was also necessary to operate the system through an external personal computer, using the teach pendant of the robot only for a few special routines not performed in every day normal operations. The robots work as slaves to that central PC, where all the parameterization is performed. The PC also monitors the operation, being of guidance when something wrong happens. The operator is able to solve problems from the PC. There is one PC for each robot, which was done for practical reasons, but it is not a requirement.

A client-server software architecture was adopted. The robot controller software works as a server, exposing to the client a collection of services that constitute its basic functionality. A collection of services was designed to fulfill all the tasks required of the system, so that they could be called from the PC (Figure 5.12). The software architecture used in this work was presented in detail elsewhere [2 -8] (see also Section 3.2), and is distributed using a client-server model based on software components (*ActiveX controls*) [10-11] developed to handle equipment functionality.

The system is completely operated using a graphical panel running on the PC, built using the above mentioned *ActiveX controls* in *Visual C++ .NET 2003* [12]. When the system is started, the operator needs only to specify what product model will be used in each pallet, and if first pallets are fully assembled. This need is only for the de-palletizing subsystem, because there is no identification on the pieces (they are coming from the high temperature oven). On the palletizing subsystem, there is no need to specify the model, because the pieces carry barcodes, inserted by the inspecting operator, that are used by the subsystem with the help of barcode readers.

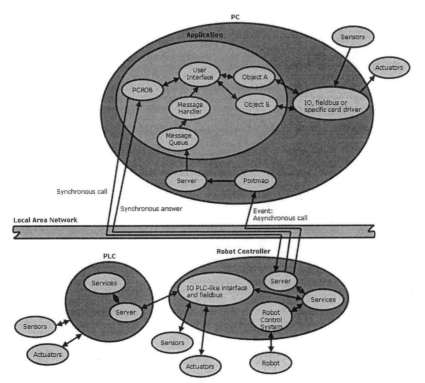

Figure 5.12 Software architecture used in this example

Sometimes, there are some non-fully assembled input-pallets on the shop floor that need to be introduced into the system. To do that, the software allows the operator to specify the position and level of the first piece. That is, however, only possible on the first pallet, because the system resets definitions to the next pallets to avoid accidents, i.e., proceeding pallets are assumed to be fully assembled. The same happens with output-pallets, since the system must be able to fill a pallet not completely filled on the last production cycle for that model.

5.3.2.1 Basic Functioning of the De-palletizing System

When the operator commands "*automatic mode*" the robot approaches the selected input-pallet in the direction of the actual piece, searchers the piece border using optical sensors placed on the gripper, and fetches the ceramic piece. After that, the robot places the piece in the first available inspection line, alternating inspection lines if they are both available, i.e., the robot tries to alternate between them, but if the selected one is not available then the other is used if available. If both inspection lines are occupied, the robot waits for the first to become available.

Figure 5.13 shows the interface used by the operator to command the system and monitor production. It shows the commands available, and the online production data that enables operators to follow production. All commands and events are logged into a log file, so that production managers can use it for production monitoring, planning, debugging, etc. The system also uses a database, organized in function of the model number, where all the data related to each model is stored. That data includes type of piece, dimensions, height where the gripper should grab the piece, average position of the first piece of the pallet, height of the pallet, and so on. Accessing and updating the database is done in "*manual mode*", selected in the PC interface.

There is a "*teaching*" option that enables operators to introduce new models and parameterize the database for that model, where a "*teach by showing*" strategy is used. When that option is commanded, the robot pre-positions near the input-pallet and the operator can jog the robot using function keys to the desired position/orientation. Basically the de-palletizing operation is preformed step-by-step and the necessary parameters acquired in the process, asking the operator to correct and acknowledge when necessary. The operator is asked to enter only the "*model number*" to teach, the height, and the width of the piece. The rest is automatic. After finishing this routine the model is introduced into the database, and the system can then work with that model number.

The system is able to check for errors such as: wrong pallet for model, presence of pallet, model not known, no piece in place, wrong level, etc. Proper warnings are sent to the PC for operator information, and displayed using software icons and short messages.

5.3.2.2 Basic Functioning of the Palletizing System

A similar approach was used for the palletizing operation. Two inspection lines are also used, with the robot trying to alternate between them. But the first available piece is removed not slowing down production. A similar approach to the one used in the de-palletizing sub-system is used to "*teach*" models to the robot. Also, the system identifies the model number from the piece barcode when "*automatic mode*" is commanded, fetches the piece, and inserts it in the pallet compatible for that model. The operator is able to select what pallet to use first, how many pieces are already there, and how many pieces it should carry (Figure 5.14). Do to the required dimensions of the output-pallets, the robot was placed on the top of a linear axis, controlled by the robot control system (robot external axis), so that a wider area could be reached. The system is also able to check for errors such as: wrong pallet for model, presence of pallet, model not known, no piece in place, etc. Proper warnings are sent to the PC for operator information, and displayed using software icons and short messages.

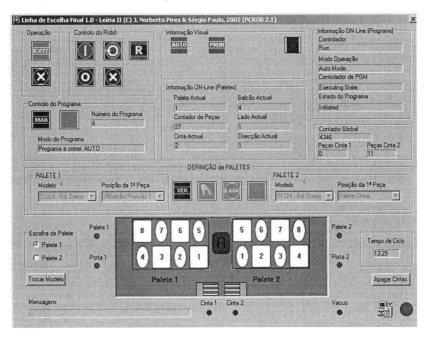

Figure 5.13 Example of an interface used by operators (de-palletizing system)

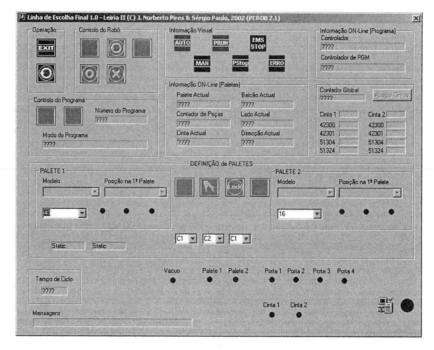

Figure 5.14 Example of an interface used by operators (palletizing system)

5.3.3 Operational Results and Discussion

The system achieved the required operational results and is flexible enough when introducing new models. Currently it works two shifts a day, almost autonomously, making around 1400 pieces per shift (one shift is seven and-a-half hours). Operators adapted easily to the system, and found the touch-screen interface easy to use.

The company improved production quality and reduced production costs: fewer operators are needed and production is more efficient (more pieces are handled a day). This can be demonstrated by operational results, and also by the fact that new systems followed this one to handle other type of pieces and other types of operations, creating a strong connection between our university and this company.

A few innovations and technology transfers were successfully introduced with this project and others are ongoing with the same company [2-5]. An interesting human-machine interface for robotic manufacturing cells was introduced with good results [2-5]. The solution has been developed from the scratch using *Visual C++ .NET 2003*, constituting a software platform that can be used with other applications. Experience with operators is positive, showing that they adapted well and really enjoy using it. Nevertheless, new developments are necessary so as to guide operators and reduce operator training. This means that advanced help should be available to guide the operator when inconsistencies are detected. Such inconsistencies include, for example:

- Commanding *"automatic mode"* without reviewing the pallets parameterization. That could be correct in some situations and consequently, allowed. At the moment only a visible warning is issued, but in the future only some sequence of operations will allow *"automatic mode"*
- Ordering a *"RSTART"*, i.e., proceed with current configuration and from the same program position, after a system stop due to an error or operator manual stop. Actually this situation is permitted, after confirming the password, because we still rely on operator training and judgment. Nevertheless, in the future, operators should be guided to follow a certain procedure, reviewing actual status, so as to avoid mistakes. This can certainly be done, for example, using an inference mechanism based on *fuzzy logic*

The two presented situations are good examples of needed future developments. For a certain industrial robotic cell characterized by a set of available operations, a collection of routes should be defined considering all possible operational situations. Consequently, an operator can command the robotic cell if he follows one of those routes. This will increase safety, avoid errors, and improve efficiency. At the moment, critical operations require operator confirmation with a password, and visible warnings issued to the screen.

Another interesting innovation was the utilization of a client-server architecture, explained elsewhere [2-5] (see Section 3.2), developed by the first author, to be used with robotic cells. Using this architecture implies the clear intention to distribute functions to all *"intelligent"* components of the robotic cell, leaving to the central PC (the client) the tasks of making the service request calls, properly parameterized, and displaying system information to the user. The PC is the user's commanding interface, and his window to the system.

5.4 Handling Production Changes Online

In this section, the problem of handling production variations online, i.e., during actual production, is addressed. These variations may occur when it isn't possible to exactly guarantee working conditions during a production cycle or between two consecutive cycles. These variations are common in some types of industries, like the glass and ceramic industry, where the products may change slightly during the production cycle. Also, these industries are multi-model industries in which the production equipment is required to handle several different models of products that have their own production requirements. Since it is common to have two or more different model campaigns during a working day, it should be possible to easily parameterize the production system when a new campaign is started.

Consequently, this section uses a highly efficient robotic palletizing system, developed for a partner company, to introduce and explain how these problems may be addressed. It includes details about practical implementation, along with a discussion of options and obtained operational results, which show the system to be a good example of human-machine cooperation.

As is common in several industries, the intermediate products need to be palletized in several stages of the production cycle, to circulate between manufacturing cells, be sold to other companies (white-line or undifferentiated products) that finish the production cycle adding their own characteristics, or to be stored inside the company in accordance with the defined production planning and company needs.

This case, the products are several models of automobile side-window glass. The palletizing system is placed after the glass cutting and washing cells. The obtained pallets are to be used in the manufacturing line that introduces the characteristic curvature of the glass. This line, which includes a high-temperature oven and an incurving system, is shared by all models of side-window glass produced by the company, which makes the task of automatically feeding the line from all cutting and washing lines very difficult to manage. Consequently, the glass is palletized using a robot manipulator and de-palletized near the incurving manufacturing line by another robot. This enables the company to handle all types of models in a very simple and efficient way.

5.4.1 Robotic Palletizing System

The system used in this example was developed to pick side-window glass from the production line and palletize it into pre-configured pallets. The system, depicted in Figures 5.15 and 5.16, is made of the following components:

- An industrial robot ABB IRB 4400, equipped with the 2002 version of the ABB S4C+ robot controller
- A PLC Siemens S7-300, to control all the systems peripheral to the robot.
- A centering system, placed on the production line, that guarantees that glasses are centered and placed in a known position before being picked by the robot
- A pneumatic gripper with retractile contact sensors and suction cups, capable of picking glasses and measuring the pallet characteristics
- A rotating system that supports two pallets, ensuring that a new empty pallet is immediately fed into the system when the previous one is full
- A computer for supervision and control, and for implementing also the human-machine interface

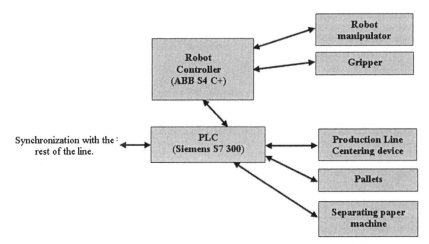

Figure 5.15 Components of the palletizing system for the automobile industry

The cycle executed by the system (Figure 5.17) is composed of the following principal tasks:

5.4.1.1 Identify Empty Pallets and Measure Parameters of an Empty Pallet
An empty pallet needs verification to measure the following pallet parameters: angle of the back of the pallet with the vertical axis, angle of the base of the pallet with the horizontal axis, height of the base of the pallet relative to the robot world reference system, and the pallet dimension. These four values change significantly from pallet to pallet and need to be obtained each time an empty pallet is introduced in the system. This task is fundamental for the success of the palletizing task, because it enables the system to place the glass always in the same conditions: at the same height relative to the pallet base and at the same distance from the previous glass. This avoids adding defects to the glass, namely small scratches on the surface of the glass (due to slipping between consecutive glasses), or on the edges that contact with the surface of the pallets (due to releasing the glass more than 1-2mm high from the surface of the pallet).

Any empty pallet needs to be measured for the above mentioned parameters that will be used during the palletizing process using that pallet. Every time the rotating base introduces a new pallet, optical sensors, placed behind the back of the pallet, detect if the pallet is empty and trigger the measuring process.

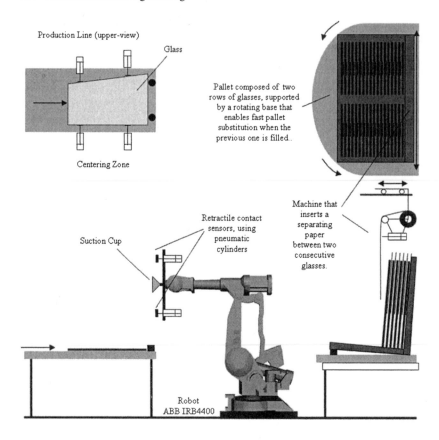

Production Line (upper-view)

Glass

Pallet composed of two rows of glasses, supported by a rotating base that enables fast pallet substitution when the previous one is filled..

Centering Zone

Machine that inserts a separating paper between two consecutive glasses.

Retractile contact sensors, using pneumatic cylinders

Suction Cup

Robot
ABB IRB4400

Figure 5.16 General view of the palletizing cell

5.4.1.2 Pick a Glass from the Production Line

After getting information from the PLC that there is a glass available in the production line, properly centered and in position, the robot is commanded to pick the glass from the predefined picking position (based on the glass model) and take it to a position near the entrance of the pallet.

5.4.1.3 Palletize the Glass

The glass must be placed in the row in use, taking into consideration the number of glasses already palletized and the pallet parameters. This operation means also knowing the thickness of the glass in a way to maintain the same palletizing conditions for all glasses. At the end, when a pallet is full, the robot signals the PLC that the pallet is full and places itself in a non-collision situation with the pallet, enabling the PLC to start the rotating motion that will exchange the pallets (Figure 5.18).

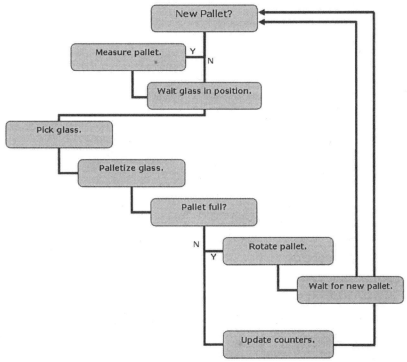

Figure 5.17 Palletizing cycle executed by the robot in automatic mode

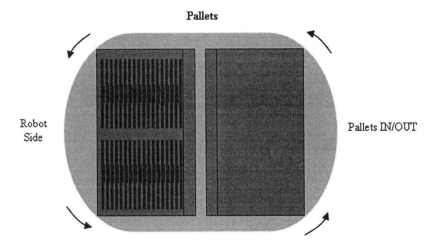

5.4.2 System Software

Considering that the above presented system was developed to work with several models of glass (up to 128 different models), that require their own configuration in the tasks of picking and palletizing each glass, i.e., these tasks are model dependent, the operating software should explore the teach-pendant capabilities in the phase of teaching a new glass model to the system. Consequently, the software was designed to have two operating modes: manual and automatic.

Manual Mode – In this mode, all subsystem testing and maintenance routines are allowed (Figure 5.19). The user is also allowed to teach a new model to the system. This means that the robot will follow pre-determined motions, asking the operator to adjust positions using function keys. In the process, the software acquires the necessary data to completely handle that model of glass. In this mode, the production line is not operational, because production is deactivated. The robot is commanded from the robot teach-pendant (or console), using local software designed to assist the selected functions. For practical reasons, this "*manual mode*" software will not be explained further here.

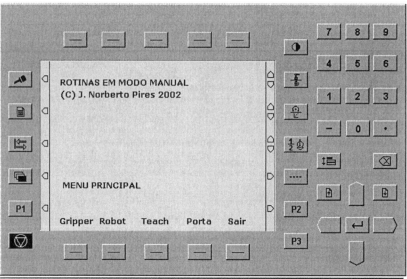

Figure 5.19 Pallet main shell presented to the user in "Manual Mode" on the robot console (original software with Portuguese interface)

Automatic Mode – The production line is placed in automatic mode and the robot should follow the cycle presented briefly in Figure 5.17. The robot uses the definitions stored in the database to handle the model selected by the operator, using the parameterizations he chooses.

The software developed to interface with the operator runs on a remote computer, connected to the robot controller by *Ethernet*. The software was developed in *Visual C++ .NET 2003* [12], using an ActiveX control [10-11] designed by the author to work with industrial robots [2-5] (see Section 3.2). The shell presented in Figure 5.20 is the operator interface to the system.

To initiate the system, the user must run the robot program using the operator interface. A *"start_program"* remote procedure call (RPC) [9] is issued, launching a computer program that implements a collection of services that can be requested from the PC using RPCs. After being initiated, the robot program waits for the selection of the operating mode, i.e., waits the user to command *"Automatic Mode"*, where the robot is controlled by the system PLC using the parameterization selected by the user, or *"Manual Mode"* where the robot is commanded from the robot teach-pendant. Both operating modes may be considered as services that the robot (*server*) offers to the PC/operator (*client*). During the *"mode selection state"*, where the robot waits for the user to select the operating mode, it is possible to access the system database where the definitions for each model are stored. Access to database is not allowed in any other situation, for safety reasons. Consequently, before selecting the operating mode, the user should select the model he wants to produce and parameterize the production: thickness of the model, number of pieces per row and per pallet, and the dimension of the glass. The thickness and dimension of the glass are characteristics of the model registered in the database, and consequently are not to be changed by the user. A password is required to change them.

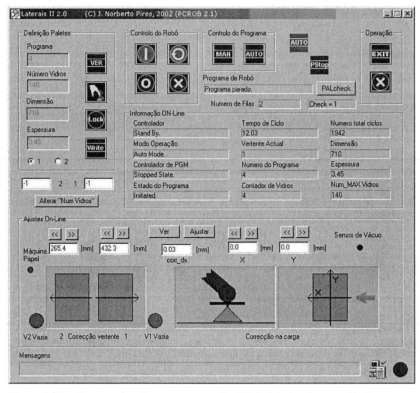

Figure 5.20 – Operator interface running on the PC (original software with Portuguese interface)

Using the interface presented in Figure 5.20, the operator is allowed to command three types of operations: Access the glass model definition database, control the robot program, and online monitoring.

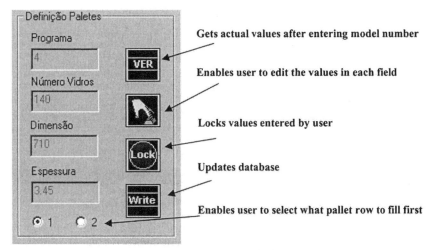

Figure 5.21 Accessing the database

Figure 5.21 shows the place where the user can change the glass model definition database. This operation is only possible, nevertheless, when the robot is waiting for operating mode selection. This procedure was implemented done for safety reasons, in a way to avoid corrupting the working database.

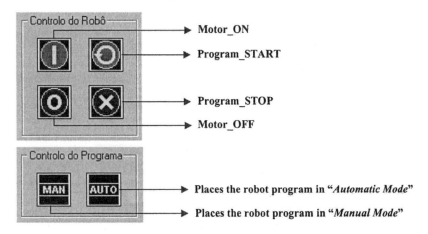

Example: Manual mode commanding routine (Visual C++ :NET 2003)

```
void CFornoDlg::Onmanual()
{float valor;
  fprintf(log,"%s - Comando de MANUAL.\n",tbuffer);
  if (m_pon.InitClient("babylon",5) >= 0)
  {valor=1236;
  nresult = m_pon.WriteNum("decision1",&valor);
  if (nresult <0) {m_log.SetWindowText("Error in the MANUAL command.");
```

```
    fprintf(log,"%s   -   Error   in   the   MANUAL   command.\n",tbuffer);erro=1;
m_erro.ShowWindow(1);}
  else m_log.SetWindowText("MANUAL command.");
  m_pon.DestroyClient();
} else
{m_log.SetWindowText("Robot didn't answer ... operation cancelled.");
 m_comms.SetIcon(AfxGetApp()->LoadIcon(IDI_smile2));
 m_erro.ShowWindow(1);
 }
}
```

Figure 5.22 Controlling the robot program

As already mentioned, commanding automatic or manual mode means accessing to a different set of functionalities. This operating mode change procedure is implemented in RAPID (ABB programming language) with the following simplified code (database access removed for simplicity):

```
WHILE never_end=FALSE DO
   WaitUntil (decision1=1235) OR (decision1=1236)\MaxTime:=1\TimeFlag:=timeout;
   IF timeout=TRUE THEN
   ENDIF
   IF (decision1=1235) THEN
      auto_mode;              ◄──────── Module that implements the "Automatic Mode"
      decision1:=0;
   ENDIF
   IF (decision1=1236) THEN
      manual_mode;            ◄──────── Module the implements the "Manual Mode"
      decision1:=0;
   ENDIF
ENDWHILE
```

5.4.3 On-line monitoring

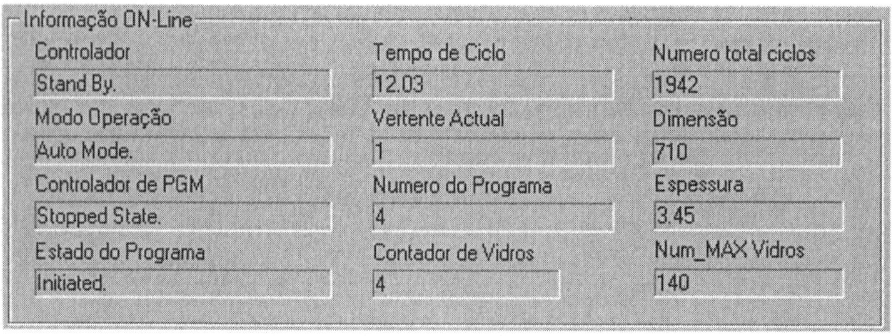

Figure 5.23 Online monitoring data

This feature (Figure 5.23) allows the user to quickly observe production data, such as: model in use, pallet row in use, number of cycles (pieces) performed since the last counter erase, number of glasses palletized in the current pallet, last cycle time, robot working modes, and so on. This information is obtained directly from the robot, making monitoring calls to the relevant services. These calls are triggered by a timer interrupt routine, programmed to monitor the system in cycles of five seconds. A complete cycle, i.e., the operation of picking and palletizing a glass, takes about nine seconds, which justifies the polling monitoring option and the choice of a monitoring cycle of five seconds.

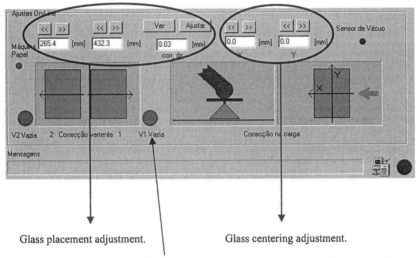

Glass placement adjustment.   Glass centering adjustment.

Note – The green and red indicators show permitted and error situations, respectively. Consequently, when a red indicator is present, the operator should interpret the warning and act accordingly.

Figure 5.24 – Adjusting online

Many times, due to operational difficulties in the production line, or centering errors, etc., it is necessary to make small adjustments in the palletizing process without stopping production. The operator may perform those adjustments using only a mouse (Figure 5.24), observe results, and correct the problem without stopping production. This type of procedure is fundamental for production environments characterized by high production rates and very tight quality control, as is the case of the automobile components industry.

Finally, another important operation under "*Automatic Mode*" is the operation of measuring the pallet parameters. That is done, as already mentioned, when a new empty pallet is introduced. This measurement must be done in every pallet, since they differ from each other significantly. Without this procedure, the palletizing process would fail. The robot is commanded to extend the precision contact sensors and use them to measure the pallet parameters. The robot uses three contact sensors, placed in the vertices of a triangle, to orient itself parallel to each surface and compute the angles around the robot's world reference system (Figure 5.25).

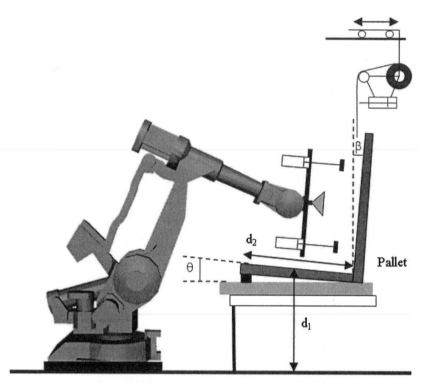

Figure 5.25 Getting pallet parameters: d_1, d_2, θ and β

The routine associated with this process is very simple and is presented below in a simplified form:

```
PROC check_pal()
    WaitUntil (divazia1=0) AND (divazia2=0)\MaxTime:=5\TimeFlag:=timeout;
    IF timeout=TRUE THEN
      TPWrite "Pallet not empty ...";
      PulseDO doerros;
      EXIT;
    ENDIF
    MoveJ pal_app,velocity,z100,toolt;
    sensores_on;
    MoveL RelTool(pal_up,0,0,250),velocity_app,fine,toolt;
```

Empty pallet??

Contact sensors in position

// Angle of the back of the pallet with the vertical axis

```
    SearchL\PStop,disen1,temp,RelTool(pal_up,0,0,500),velocity_search,toolt;
    MoveL temp,v10,fine,toolt;
    temp:=CRobT(\Tool:=tool_sen1);
    WHILE (disen2=0) AND ((disen3=0)) DO
```

```
MoveJ RelTool(temp,0,0,0\Ry:=-0.1),velocity_search,fine,tool_sen1;
temp:=CRobT(\Tool:=tool_sen1);
ENDWHILE
pal_actual:=CRobT(\Tool:=toolt);
angle1:=Abs(90-Abs(EulerZYX(\Y,pal_actual.rot)));
TPWrite "Back Angle = "\Num:=angle1;
```

// **Angle of the base of the pallet with the horizontal axis**

```
MoveJ pal_up,velocity_app,fine,toolt;
MoveJ pal_down,velocity_app,fine,toolt;
SearchL\PStop,disen1,temp,RelTool(pal_down,0,0,500),velocity_search,toolt;
MoveL temp,v10,fine,toolt;
temp:=CRobT(\Tool:=tool_sen1);
WHILE (disen2=0) AND ((disen3=0)) DO
  MoveJ RelTool(temp,0,0,0\Ry:=-0.1),velocity_search,fine,tool_sen1;
  temp:=CRobT(\Tool:=tool_sen1);
ENDWHILE
WaitTime 0.2;
temp:=CRobT(\Tool:=toolt);
angle:=Abs(EulerZYX(\Y,temp.rot));
TPWrite "Base Angle "\Num:=angle;
temp1:=RelTool(pal_actual,-(dim{modelo}/2-(pal_actual.trans.z-temp.trans.z)),0,0);
pal_actual:=temp1;
MoveJ pal_down,velocity_app,z50,toolt;
MoveJ pal_app,velocity,z100,toolt;         Height and dimension of the pallet
sensores_off;
ENDPROC
```

Retract contact sensors

5.4.4 Discussion and Results

The system (Figure 5.26) presented in this section is a good example of a flexible robotic industrial system, capable of handling any production situation. The system relies on operator command and judgment, enabling him to fully parameterize production and introduce new production models. Besides of that, the operator may also introduce adjustments and change working conditions online, without stopping production, which is a powerful tool to handle production variations and difficulties. These features were obtained just by implementing a collection of services capable of handling all the anticipated production requirements, exposing them to the remote computer (*client*) where the operator interface is implemented. In this way, production may be tailored in a very flexible way, enabling the operator to solve virtually any operational situation.

Operational results are promising:
- Operators adapted easily to the system, which is always a good result considering their average skills
- Achieved production cycle is of aboutnine seconds per glass, which is more than is required

- The pallet measuring procedure takes about 25 seconds to complete, which is compensated by the very fast cycle time. The average overhead introduced by this procedure in the cycle time is about $25/280 = 0,089 \sim 0,1s$ (taking an average number of 280 glasses per pallet), which has no meaning
- The system works 24 hours a day without any need for operator supervision

It is worthwhile to point out that this system uses a client-server architecture, explained elsewhere [2-5] (see Section 3.2), developed to be used with robotic cells. Using this architecture implies the clear intention to distribute functions to all *"intelligent"* components of the robotic cell, leaving to the central PC (*the client*) the tasks of making the service request calls, properly parameterized, and displaying system information to the user. The PC is the user's commanding interface, and his window to the system. The developed software was built from scratch and the authors didn't use any commercial software, apart from operating systems (for example, *ABB Baseware 4.0* for the industrial robots, and *Microsoft Windows 2000* with *Service Pack 4* for the PC) and developing tools (*Visual C++ .NET 2003* [12] from *Microsoft*). A port of the *SUNRPC 4.0* [9] package for *Windows NT/2000/Xp*, a free open package originally developed for *UNIX* systems, was also used. The porting effort was, nevertheless, completely done by the author.

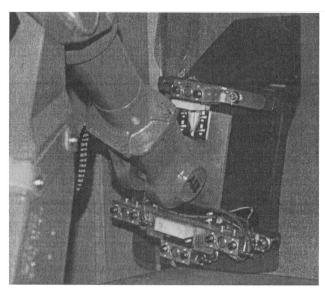

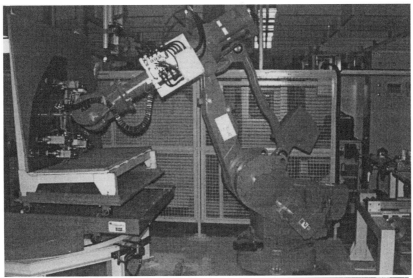

Figure 5.26 General view of the system

5.4.5 Conclusion

The system presented in this section is an implementation of a distributed software architecture developed to work with industrial robotic cells. The main objective was to be able to change production conditions online, and make adjustments to the working parameters so as to cope with production variations. The system was presented in some detail, giving special attention to the software designed to parameterize, monitor, and adjust the production setup enabling online adjustments to the working conditions. Obtained operational results demonstrate the interest of these types of systems for multi-model production environments, where high production rates and quality demands are a key factor. Finally, the obtained system is also a good example of man-machine cooperation, demonstrating the advantages of mixing human and automatic labor in actual manufacturing plants.

5.5 References

[1]   ABB Robotics, "IRB6400 User and System Manual", ABB Robotics, Vasteras, 2002.
[2]   Pires JN, Sá da Costa JMG, "Object Oriented and Distributed Approach for Programming Robotic Manufacturing Cells", IFAC Journal on Robotics and Computer Integrated Manufacturing, February 2000.
[3]   Pires, JN, "Complete Robotic Inspection Line using PC based Control, Supervision and Parameterization Software", Elsevier and IFAC Journal Robotics and Computer Integrated Manufacturing, Volume 20, N.1, 2004.

[4] Pires, JN, Paulo, S, "High-efficient de-palletizing system for the non-flat ceramic industry", Proceedings of the 2003 IEEE International Conference on Robotics and Automation, Taipei, 2003.

[5] Pires, JN, "Object-oriented and distributed programming of robotic and automation equipment", Industrial Robot, An International Journal, MCB University Press, July 2000.

[6] Pires JN, "Interfacing Robotic and Automation Equipment with Matlab", IEEE Robotics and Automation Magazine, September 2000.

[7] Pires, JN, Godinho, T, Ferreira, P, "CAD interface for automatic robot welding programming", Sensor Review Journal, MCB University Press, July 2002.

[8] Pires, JN, and Loureiro, Altino et al, "Welding Robots", IEEE Robotics and Automation Magazine, June, 2003   .

[9] Bloomer J., "Power Programming with RPC", O'Reilly & Associates, Inc., 1992.

[10] Box D., "Essential COM", Addison-Wesley, 1998

[11] Rogerson D., "Inside COM", Microsoft Press, 1997.

[12] Visual C++ .NET 2003/2005 Programmers Reference, Microsoft, 2003 (reference can be found at Microsoft's web site in the Visual C++ .NET location)

6

Final Notes

6.1 Introduction

Dear reader, I hope you had fun reading and exploring this book, because in my opinion that is a fundamental outcome of a technical book. Furthermore, a book about robotics and automation must stimulate the reader curiosity and interest to explore further on its own.

This book is a practical guide about industrial robotics and related subjects. My primary objective was to introduce you to the fantastic world of robotics and ride with you through ideas, examples, and industrial solutions showing how things can be done, what are the available alternatives and challenges. Robotics and automation is a multidisciplinary subject that calls for creativity and innovation. It poses a permanent challenge for performance and practical results and consequently is a perfect subject for inventive and dedicated people, for whom this book was written. For that reason, the book presents a considerable amount of examples and solutions, allowing readers to see, from time-to-time, the complete picture of building a robotic manufacturing system, which constitutes also an invitation to maintain the focus. That is important. Robotics is an interesting subject and people are naturally attracted by its applications and achievements. Nevertheless, due to its multidisciplinary nature, robotics is also a very demanding field requiring knowledge of physics, electronics, mechanics, computer science, and engineering. Consequently, a book in the field gains by having examples and practical implementations. That was the "*design option*" followed when planning and writing the book. You can find the code of several of the presented examples along with pictures, videos, and other material at:

http://robotics.dem.uc.pt/indrobprog

The access to the site is restricted and requires a login "*username*" and "*password*". Visit the web site for details on how to obtain a valid login. As author of this book, I'll keep the website updated so that it is a good source of information on:

- New developments
- Interesting solutions
- Interesting scientific and technical papers
- Interesting books
- Industrial trends in terms of technology

Most of these issues are related to new developments that result from R&D projects done in universities, research institutes, and companies, or in cooperation between academia and industry, resulting in technical papers, books, and new products. Robotics and automation is perhaps one of the most interesting cases of industry-academia cooperation since most of the developments require scientific, technical, and operational advances from both worlds to reach higher levels in terms of manufacturing flexibility and agility.

To be faithful to the basic *"design option"* adopted in this book, we will finish with another example. This final case is about a technical solution designed to reconfigure an old industrial robot, making it accessible through a local area network (LAN), and allowing programmers and system engineers to offer remote services to users.

6.2 Operation *"Albert"*

Albert is the name of an old robot that we acquired for our laboratory (Figure 6.1). The primary objective behind the acquisition was to obtain a nice industrial machine dedicated to teaching activities and to be included in laboratory classes of the discipline of *"Industrial Robotics"* (4th year of the Mechanical Engineering course). *Albert* worked for a few years in industry doing several types of tasks: manipulation, gluing, and labeling. After retiring from industry it is now starting a promising career in academia. Technically, *Albert* is an anthropomorphic robot manipulator (from 1992, build year) manufactured by *ABB Robotics* (model IRB1500) and equipped with an ABB S3 robot controller [1], i.e., it is a robot from 1992 but carrying technology from the mid eighties. Consequently, it is a rather old system with the following basic characteristics:

- Anthropomorphic manipulator (model ABB IRB1500): 5kg of payload, 6 axis, 0.1mm of repeatability and a fairly interesting workspace area (~1400mm)
- ABB S3 robot controller: This is the main disadvantage of *Albert*, since the S3 system is old and not carrying much of the interfaces required by actual industrial manufacturing systems. The controller is programmed using the programming language *ARLA* and has 16 digital inputs, 16

digital outputs, a serial port for data communication, and a very basic teach pendant (Figure 6.2).

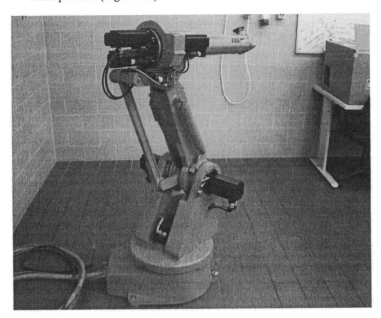

Figure 6.1 *Albert* is an ABB IRB1500 manipulator

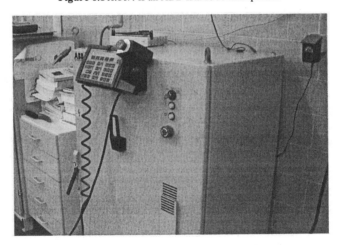

Figure 6.2 S3 robot controller

Consequently, this is mechanically a very interesting machine, very similar to its successor, the IRB1400 model. In fact, they share the same wrist design, which gives to the arm an excellent maneuverability. Nevertheless, because it is an old system with very deficient communication interfaces, without any LAN interface, an old programming language (although sufficiently powerful) and a very basic user interface, *Albert* needs to be upgraded to be useful for teaching and training tasks.

Figure 6.3 S3 cabinet with the extra hardware

To provide the system with a LAN interface, and the ability to offer programmed services to remote clients, while keeping the available system functionalities, the following actions were performed to upgrade the old *Albert* (Figure 6.3):

- A PLC (*S7-266* from *Siemens*) was added to the system, connected to the robot using the IO digital boards available in the S3 system. Consequently, a very simple parallel interface was added to transfer data between the PLC and the robot controller
- An *Ethernet* board (*CP 243-1* from *Siemens*) was also added to the system, connected to the PLC, to enable the system to interface with the LAN available in the laboratory. Consequently, remote users interface with the robot controller through the PLC, which means that a basic data protocol must be defined to exchange information between remote users and the running robot programs. That is a very simple task and was already used in Chapter 3
- An extra IO module was also added to the PLC to provide a supplemental set of IO digital line to use with applications.

The PLC is accessed using the *Ethernet* board and a simple UDP messaging system. To simplify the access, we used the *Siemens S7-200 OPC Data Access (OPC DA) Server* for the S7-200 (a server that is part of the *Siemens S7-200 PC Access* package). This server provides a means to access the PLC memory allowing the user to execute read/write operations on the entire PLC memory spaces (includes program variables, IO variables, special memory bits, etc.).

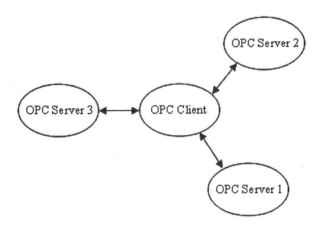

Figure 6.4 OPC client-server connection

Basically, OPC (*OLE for Process Automation*) [2, 3] was designed to allow client applications to access data from *shop floor* devices in a consistent and transparent way. Therefore, the OPC client applications interface with software modules (the OPC servers) and not with the hardware directly. This means that they rely on software components provided by the hardware manufacturer to efficiently access and explore the hardware features. Consequently, changes and hardware upgrades will not affect the user applications.

With OPC, whose specifications [3] include a set of *custom COM interfaces* [4] (used when building client applications) and a collection of *OLE automation interfaces* [5] to support clients built using high-level languages and applications (*Visual Basic* and *Excel*, for example), users can take advantage of the nice features of DCOM to facilitate client access to the system features. An OPC client can connect to OPC servers provided by any vendor that followed the OPC specification [3] (Figure 6.4).

Basically there are three types of OPC servers [2, 6]:

1. *OPC Data Access Servers (OPC DA Servers)* – This type of server is used to offer read/write data services to the client application. OPC DA servers constitute a powerful and efficient way to access automation and process control devices

2. *OPC Alarm and Event Handling Servers (OPC AE)* – This type of server is used to implement alarm and event notifications services to be used with client applications

3. *OPC Historical Data Access Servers (OPC HDA)* – This type of server is used to access (read/write) data from an historian engine

In this project to upgrade and reconfigure *Albert* an OPC DA server [7] is used to access the PLC. An OPC DA client application designed to access the PLC resources needs to deal with three types of objects:

1. *OPC DA Servers* – maintains information about the server and operates as a group container

2. *OPC DA Groups* – provides the mechanisms for containing and organizing items. Every OPC group has a particular update rate that must be set by the OPC client

3. *OPC DA Items* – the items are the real connections to the system resources. An item could represent a bit (like a memory bit or IO digital signal, etc.), a byte, a word, etc

Consequently, to access data from the hardware resource through the OPC server, the client should follow the following procedure:

- Connect to the OPC server
- Create an OPC group to perform synchronous reads/write operations
- Add the necessary items to the group
- Monitor the actual state of the items, or make asynchronous read/write operations

With *Albert,* twelve digital IO inputs and twelve digital IO outputs are used as data bus for robot-PLC communication. Some of those IO lines will be use to control the information flow between the robot and PLC. The remaining four digital inputs and four digital outputs will be used for special operations (Table 6.1).

To demonstrate how this can be used to command *Albert* from a remote PC, consider that the robot "knows" five positions, which are available for user request. The idea is to build a simple OPC client application to set up an OPC connection to the *Siemens S7-200 OPC Server*, and implement the necessary actions to command the robot to move to the user-selected positions.

Table 6.1 IO assignment for robot-PLC communication

Robot	PLC	Description
DI1 to DI12	Q0.0 to Q1.3	*Data IN*
DO1 to DO12	I0.0 to I1.3	*Data OUT*
DI13	Q1.4	*Motor_ON*
DI14	Q1.5	*Motor_OFF*
DI15	Q1.6	*Program_RUN*
DI16	Q1.7	*Program_STOP*
DO13	I1.4	*Motor_State*
DO14	I1.5	*Program_State*
DO15	I1.6	*System_State*
DO16	I1.7	*Emergency_State*

With that objective in mind, the following items were defined in the OPC server (Figure 6.5):

> **q0** – byte that contains the digital outputs Q0.0 to Q0.7
> **q1** - byte that contains the digital outputs Q1.0 to Q1.7
> **i0** - byte that contains the digital inputs i0.0 to i0.7
> **i1** - byte that contains the digital inputs i1.0 to i1.7

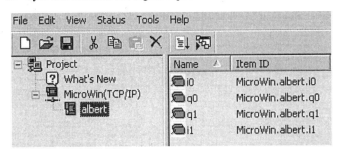

Figure 6.5 Items defined in the OPC server for this simple example

To implement the possibility of moving the robot using the OPC server, the following sequence is adopted:

- The robot waits for Q0.7 = DI8 = 1; means that a valid command is ready
- The commanded position is specified through bits Q0.0 (DI1) to Q0.4 (DI5), i.e., Q0.0 (DI1) is associated with P1, Q0.1 (DI2) with P2, ..., Q0.4 (DI5) with P5
- The robot program jumps to "MOVE P1" routine and acknowledges the received command by making DO8 = I0.7 = 1
- The commanding PC should confirm the motion just by making q0 = DI1-DI8 = 0
- Robot makes DO8 = I0.7 = 0 and moves to the commanded position.
- Robot program jumps to the beginning and waits for a new command

Consequently, the program running on the robot controller (coded using *ARLA*) looks like the generic code presented in Figure 6.6.

```
while never_end;
   wait DI8 = 1;
   switch (byte DI1-DI8)
      case 1: DO8 = 1; wait (word DI1-DI8) = 0; DO8 = 0; Move P1;
      case 2: DO8 = 1; wait (word DI1-DI8) = 0; DO8 = 0; Move P2;
      case 4: DO8 = 1; wait (word DI1-DI8) = 0; DO8 = 0; Move P3;
      case 8: DO8 = 1; wait (word DI1-DI8) = 0; DO8 = 0; Move P4;
      case 16: DO8 = 1; wait (word DI1-DI8) = 0; DO8 = 0; Move P5;
   endswitch;
endwhile;
```

Figure 6.6 – Generic code running on *Albert's* controller

The OPC client application designed to connect to the OPC server, monitor the selected items and interface with the PLC (and through it to the robot controller) is represented in Figure 6.7.

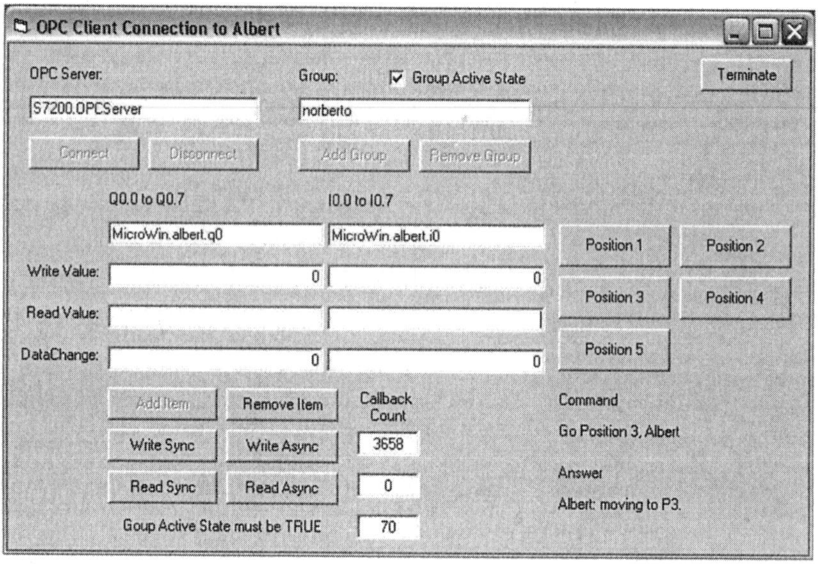

Figure 6.7 OPC client application designed to command the robot

The client application creates a group named "*norberto*" and enables the user to add the items of interest. In this, case the selected items are *Microwin.albert.q0* and *Microwin.albert.i0*. The default group updated rate is 100ms.

When a command is selected (using the software buttons "*Position 1*" to "*Position 5*"), the client application follows the above sequence just by monitoring the robot response (through the PLC interface), and acting accordingly. Figure 6.8 reveals the code associated with the action of commanding the robot to move to P1.

```
Private Sub p1_Click()
  If txtChangeVal(1).Text = "0" Then
    txtWriteVal1.Text = "129"        ←————— Command valid + MOVE to P1
    lp1 = 1
    cmdWriteAsync           ←————————————— Call to WriteAsynchronous
    cmd_sent.Caption = "Go Position 1, Albert"
  Else
    cmd_sent.Caption = "Albert: I'm not ready!"
  End If
End Sub

Private Sub Timer1_Timer()
  If (lp1 = 1) Then
    If (txtChangeVal(1).Text) = "128" Then ←┬— Robot received the command
      txtWriteVal1.Text = "0"
      cmdWriteAsync        ←—————————┤    Call to WriteAsynchronous
      lp1 = 0
      answer.Caption = "Albert: moving to P1."
    End If
  End If
  ...
  If (lp5 = 1) Then
    If (txtChangeVal(1).Text) = "128" Then  ——┘
      txtWriteVal1.Text = "0"
      cmdWriteAsync        ←—————————— Call to WriteAsynchronous
      lp2 = 0
      answer.Caption = "Albert: moving to P5."
    End If
  End If
End Sub
```

Figure 6.8 Code associated with the command action move to P1

This example shows clearly the usefulness of the updated *Albert* for teaching and training tasks. In the update process a PLC was added to the robot controller cabinet, including an extra IO board and an *Ethernet* card (on the PLC bus), which can work in parallel with the application running on the robot controller. These new features can be explored when building applications, and since the user needs to deal with the robot controller software, the PLC software, and the protocol to manage the robot-PLC communication (as shown in the presented example), it is fair to say that the new *Albert* constitutes a very nice platform to learn about robotics and automation.

6.2.1 And "*Albert*" Speaks

From the material presented in Chapter 4, the task of adding a speech interface to *Albert* is straightforward. Nevertheless, it will be done in this section, step-by-step, because in the process a few details about the human-robot interface will be further clarified. For simplicity, we'll use the same setup presented above.

The first thing to decide is the structure of the voice commands. The best option is the "*command and control mode*" (see Section 4.2.3) because it is more adapted to industrial situations that require a clear and safe identification of commands. With this operation mode, the software needs to identify the sequence of words and strings that compose the command, and generate the appropriate action to the robot controller. Consequently the selected command structure is

name_of_machine command parameters

where "*name_of_machine*" is the name attributed to the machine (in our case "*Albert*" or "*robot*"), "*command*" is a word identifying the command and "*parameters*" are words or strings identifying the parameters associated with the particular command.

In the presented example, there are four commands available:

"*hello*" – enables the user to query if the interface is available
"*initiate*" – initiates the speech interface
"*terminate*" – suspends the speech interface
"*move*" – commands the robot to move to a position

These commands are associated to the machine "*Albert*" (or "*robot*"), which means that they are associated with the pre-command string "*Albert*" (or "*robot*").

The next step is to write the above defined grammar using a standard format that can be understood by our software. There are two ways to achieve that:

- Include grammar specific instructions in the body of the software (hard-coded grammar). This means that any change in the grammar structure, or a simple update in the command list, requires another compilation of the application software.
- Specify the grammar using XML files. This is straightforward and flexible to changes and updates.

In the presented example, an XML file is used to specify the grammar (Figure 6.9). Since we use English and Portuguese recognizers, two XML grammars were built to allow the user to select the language. The application reads the grammar from the XML file, selects the recognizer to use based on the language ID tag, commits the rules, and handles the recognition events. When a certain rule is identified, an event is fired by the recognition engine and catch by our application that executes the appropriate actions (Figure 6.10).

```
<GRAMMAR LANGID="409">
 <DEFINE>
  <ID NAME="test" VAL="1"/>
  <ID NAME="move" VAL="2"/>
  <ID NAME="position" VAL="3"/>
  <ID NAME="init" VAL="4"/>
 </DEFINE>
 <RULE NAME="ROOT" TOPLEVEL="ACTIVE">
  <L>
   <P>albert</P>
   <P>robot</P>
  </L>
  <RULEREF PROPNAME="move" PROPID="move" NAME="move"/>
  <P>to</P>
  <RULEREF PROPNAME="position" PROPID="position" NAME="position"/>
  <O>please</O>
 </RULE>
 <RULE NAME="START" TOPLEVEL="ACTIVE">
  <L>
   <P>albert</P>
   <P>robot</P>
  </L>
  <RULEREF PROPNAME="init" PROPID="init" NAME="init"/>
  <O>please</O>
 </RULE>
 <RULE NAME="move">
  <LN PROPNAME="move" PROPID="move">
   <PN VAL="1">move</PN>
   <PN VAL="2">go</PN>
  </LN>
 </RULE>
 <RULE NAME="init">
  <LN PROPNAME="init" PROPID="init">
   <PN VAL="1">initialize</PN>
   <PN VAL="2">terminate</PN>
   <PN VAL="3">hello</PN>
  </LN>
 </RULE>
 <RULE NAME="position">
  <LN PROPNAME="position" PROPID="position">
   <PN VAL="1">position one</PN>
   <PN VAL="2">position two</PN>
   <PN VAL="3">position three</PN>
   <PN VAL="4">position four</PN>
   <PN VAL="5">position five</PN>
  </LN>
 </RULE>
</GRAMMAR>
```

Figure 6.9 XML file containing the speech grammar (English version)

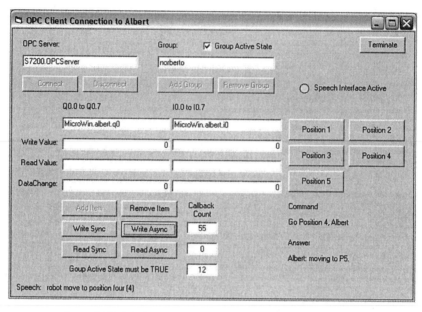

Figure 6.10 OPC client application with the speech interface included

Figure 6.11 show the code associated with the rules that command the robot to move to position one:

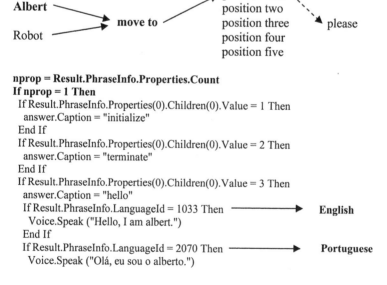

```
nprop = Result.PhraseInfo.Properties.Count
If nprop = 1 Then
  If Result.PhraseInfo.Properties(0).Children(0).Value = 1 Then
    answer.Caption = "initialize"
  End If
  If Result.PhraseInfo.Properties(0).Children(0).Value = 2 Then
    answer.Caption = "terminate"
  End If
  If Result.PhraseInfo.Properties(0).Children(0).Value = 3 Then
    answer.Caption = "hello"
    If Result.PhraseInfo.LanguageId = 1033 Then            English
      Voice.Speak ("Hello, I am albert.")
    End If
    If Result.PhraseInfo.LanguageId = 2070 Then            Portuguese
      Voice.Speak ("Olá, eu sou o alberto.")
```

```
      End If
      End If
   End If
   If (nprop = 2) Then
     If Result.PhraseInfo.Properties(1).Name = "position" Then
     If (Result.PhraseInfo.Properties(1).Children(0).Value = 1) Then
       speech_out.Caption = speech_out.Caption + " (1)"
       If Result.PhraseInfo.LanguageId = 1033 Then
         Voice.Speak ("Position one, master.")
       End If
       If Result.PhraseInfo.LanguageId = 2070 Then
         Voice.Speak ("Posição um, mestre.")
       End If
       p1_Click ─────────────►  Routine that commands the robot to move to P1
     End If
   (...)
```

Figure 6.11 Visual Basic code associated with handling speech events: aspects related with the "move to position" command

When an event is received, the application needs to query the speech API for the property that was identified, and take the appropriate actions based on the returned values. It's a straightforward procedure based on the selected command structure defined in the XML file containing the speech grammar.

With this example, I finish this book. My sincere hope is that it could constitute a nice and useful resource of information and inspiration, but also a *"platform"* to stimulate your curiosity to proceed further in the area.

Because... Robotics is Fun!

6.3 References

[1]   ABB Robotics, "IRB1500 Users and Systems Manual", ABB Robotics, Vasteras, 1992.
[2]   Iwanitz, F., Lange, J., "OPC, Fundamentals, Implementation and Application", Huthig, 2nd edition, 2002.
[3]   The OPC Foundation, http://www.opcfoundation.org
[4]   Box D., "*Essential COM*", Addison-Wesley, 1998
[5]   Rogerson D., "*Inside COM*", Microsoft Press, 1997.
[6]   OPC Foundation, "OPC Overview", Version 1, OPC Foundation, 1998.
[7]   Siemens Automation, "S7-2000 PC Access Users Manual", Siemens, 2005.

Index

Printed in the United States